OMNI'S
CONTINUUM

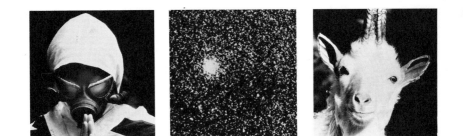

OMNI'S CONTINUUM

DRAMATIC PHENOMENA
FROM THE NEW FRONTIERS OF SCIENCE

EDITED BY DICK TERESI

Little, Brown and Company   Boston/Toronto
An Omni Press Book

This book is dedicated to the person—whoever
and wherever he or she may be—who invented silver ink.

MV

*Published simultaneously in Canada
by Little, Brown & Company (Canada) Limited*

Printed in the United States of America

PREFACE

Sir Walter Scott laughed when he heard the idea of the gaslight. Dwight D. Eisenhower derided the importance of orbiting space satellites. "The Russians have put a small ball up in the air," he said about Sputnik. "That does not raise my apprehensions one iota." Lee DeForest, inventor of the electron tube and the "father of electronics" said, in 1926, that "television may be feasible, yet commercially and financially I consider it an impossibility. . . ." By 1957 he no longer doubted TV, but instead attacked the notion of men landing on the moon, calling it a "wild dream worthy of Jules Verne."

H.G. Wells predicted that submarines could do no more than drown their crews at sea. Wilbur Wright confessed that he told Orville in 1901 that it would be fifty years before man would fly. And even Thomas Edison said, "They never will try to steal the phonograph. It is not of any commercial value."

Which just proves that even geniuses can become mired down in the miasma of pessimism.

This is not a book about pessimism. Nor is it a book of unbridled optimism. It is a book of facts, and the significance of those facts. Because if you know the truth—what scientists *today* are discovering, inventing and exploring— you will see more clearly what the future holds.

When *Omni* was founded in 1978, the great seers of publishing predicted the magazine would fail immediately, that the public would not care about science. As it turned out, the magazine has become an impressive success. And the most popular department of *Omni* is not fiction or speculation, but a monthly, eight-page section of news stories that covers discoveries in science on a continuing basis. Not surprisingly, it is called Continuum.

You will discover within these Continuum stories that some of your wildest dreams not only are very possible, but may have already happened.

—Dick Teresi

ACKNOWLEDGMENTS

So many people have contributed to Continuum over the years, and helped shape its content, style, and visual appeal, that it would be impossible to name them all here. Below, however, is a partial list of those who contributed to this particular volume.

Editor: Dick Teresi

Design Director: Frank DeVino

Book and Cover Designer: Paul Slutsky

Cover Photographer: Dan Morrill

Photo Researcher: Rod Rodriguez

Photographs courtesy of United Press International; John Muth (p. 27); NASA (p. 87); Pat Hill (p. 168); Bailliere Tindall—*Science Digest* (p. 166); Steve Costillo (p. 245)

Senior writers: Patrice Adcroft, Douglas Colligan, Michael Edelhart, Jeff Hecht, Judith Hooper, Allan Maurer, Ken Rose, and Pamela Weintraub.

Other contributors: Ellen Bilgore, Alton Blakeslee, Jane Bosveld, Robert Brody, David Clutterbuck, Margaret Cottey, David Cohen, Leonard David, Owen Davies, Bernard Dixon, Sandra Dorr, Nick Engler, Peter Evans, Barbara Ford, Kendrick Frazier, Al Furst, Susan Gertman, Don Hinrichsen, Phoebe Hoban, Carol A. Johmann, Robert Kall, Anne Klein, John Kogut, Tom Kovach, Susan Lang, Chris Larson, Harry Lebelson, Madeleine Lebwohl, Richard Levine, Lisa Levinson, Irving Lieberman, Lajos Matos, Kathleen McAuliffe, Marc McCutcheon, Dave McNary, Eric Mishara, Sy Montgomery, John Newell, James Randi, Caroline Rob, Peter Rondinone, Margaret Sachs, Howard Smallowitz, Ivor Smullen, John Stansell, William Stuckey, Mark Teich, Nicholas Timmins, Anthony Tucker, Robert Vogel.

Special thanks must also go to Bob Guccione and Kathy Keeton, who originally conceived of and created Continuum, and to Barry Lippman, John Evans, Beverly Nerenberg, Marcia Potash, Fran Lunzer, Michael Weinglass, Tom Stinson and Murray Cox for their efforts on this book. *Omni* also wishes to cite Hildegard Kron and Margaret Richichi for their ongoing visual contributions to Continuum.

CONTENTS

		PREFACE	
Chapter	1:	HEALTH AND MEDICINE	3
Chapter	2:	ANIMALS AND WILDLIFE	45
Chapter	3:	STARS AND SPACE	67
Chapter	4:	BEHAVIOR AND THE MIND	91
Chapter	5:	ENERGY	123
Chapter	6:	TECHNOLOGY	137
Chapter	7:	PHENOMENA	165
Chapter	8:	INVENTIONS AND INNOVATIONS	189
Chapter	9:	ENVIRONMENT	217
Chapter	10:	THE PARANORMAL (AND OTHER WEIRD THINGS)	231
		SUBJECT INDEX	246

OMNI'S CONTINUUM

HEALTH
AND MEDICINE

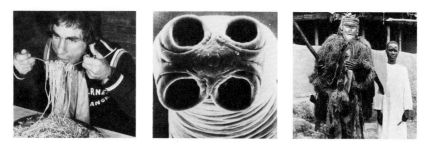

CHAPTER 1

Diet foods of the
future may turn out to be pasta,
weenies, cookies, and
other foods high in carbohydrates.

BROWN-FAT DIET Can a diet of high-carbohydrate foods, such as pasta and cookies, help a person to lose weight? The idea may not be as crazy as it sounds, although dieters are advised not to trade in their cottage cheese for Twinkies—not just yet.

The secret is brown fat, which researchers may ultimately exploit to fight obesity. Unlike the familiar white fat, which is merely a storage depot for calories, brown fat *burns* calories to provide heat. Found mainly in the neck and kidney areas and between the shoulders, it gives small mammals and newborn humans vital protection against the cold.

Until recently it was believed that adult humans, with a limited need for defense against cold, have almost no brown fat at all. At the same time, a celebrated British study showed that overeating stimulates the production of brown fat.

Spurred by these findings, researchers like Elliot Danforth, of Vermont University, and Lewis Landsburg, at Beth Israel Hospital, in Boston, began investigating ways to manipulate brown-fat levels through nutrition. Danforth demonstrated that volunteers in prison who were overted gained less weight than expected, largely because their brown-fat production sped up, burning off most of the extra calories. Thin inmates had more difficulty than obese ones both in gaining weight and in maintaining their new weight. One reason: They had more brown fat.

The researchers now agree that some people are naturally obese and some are naturally slim; fat people do not produce sufficient brown fat to burn off extra calories, but skinny people produce it in superabundance.

How can fat people have more of what thin people have? What foods, if any, might speed up the production of brown fat? One answer may be carbohydrates. In the Vermont study, the test subjects were fed cafeteria diets high in carbohydrates—pasta, Spam, weenies, and cookies. The carbohydrates activated production of an especially high volume of brown fat. Still, it is not yet clear whether a high-carbohydrate diet will someday be a key to

The health secrets of plants, known for years by tribal medicine men, are now being confirmed in the lab.

human weight reduction; most of the studies thus far have been done on rats. The important testing on human subjects is just getting under way.

BARLEY JUICE, CANCER, AND AGING What tribal medicine men and herbalists have been saying for years, modern researchers are now confirming: Green plants are virtual medicine chests.

According to researchers working in California and Japan, the juice of young barley plants—and possibly other vegetables as well—contains substances capable of dramatically slowing the aging process and repairing the ravages of cancer.

In their experiments, Yoshihide Hagiwara, president of the Hagiwara Institute of Health in Japan, and Yasuo Hotta, a research biologist with the University of California, found that the addition of barley juice to damaged cells stimulates cellular DNA to repair itself.

The scientists experimented upon miotic cells in mice, exposing them to both radiation and a strong carcinogen. They then divided the cells into two sets, incubating both but adding the barley juice to only one set. The "untreated" cells fared poorly; many cells died, the research showed, while others began a slow self-repair process. But the cell group treated with the barley juice showed an enormously accelerated repair rate: 99 percent faster than normal.

But when they used the barley juice as a preventive measure on the cells before damage, their results were even more dramatic. Hotta reported a repair rate "easily 200 percent" faster than normal.

The researchers are trying to determine precisely which substance in the plants is responsible for this remarkable action. Both believe the ingredient must be present in other plants as well.

Both cancer and aging, current theories hold, result from the genes' ineffectiveness at repairing themselves. Therefore, says Hagiwara, "this report is sure to give a shock" to the medical community.

> *"Be careful about reading health books. You might die of a misprint."*
> —Mark Twain

INTERFERON LOTIONS Tests on interferon, the much-touted protein virus and cancer fighter, have not yet reached the stage at which the drug's effectiveness against cancer is known. But work on bringing the first interferon products to market is progressing.

Perhaps surprisingly, the first two interferon medications we are likely to see will be a lotion and an eyewash. The lotion, which has been tested at Memorial Sloan Kettering Cancer Center, is designed for use against severe cases of venereal herpes. Herpes is caused by a virus, and interferon has been shown to be very effective in the treatment of viral diseases. The lotion should reduce the pain and shorten the duration of recurring herpes attacks.

The eyewash, which is being developed at Novo Labs, in Wilton, Connecticut, and in Japan, will combat herpes conjunctivitis of the eye. These sores, a symptom of a second family of herpes virus, can cause great discomfort and can lead to serious damage. The interferon eyewash will reduce the swelling, lessen the pain, and quicken healing around the eye.

The products should be ready to go on the market soon. The Food and Drug Administration approves medications for external use much more quickly than it does for those taken internally, so availability of these treatments shouldn't be delayed by years of government-ordered tests.

KILLER AMOEBAS As if Legionnaire's Disease wasn't enough to worry about, now there's parasitic encephalitic meningitis (PEM), a disease that sounds like something out of a bad science fiction movie.

Caused by an amoeba that lives on the bottom of freshwater ponds and lakes, PEM is little understood and nearly always fatal. The amoeba has caused over a hundred reported deaths since its discovery in 1963, according to Dr. George Healey, parasitologist at the Federal Center for Disease Control in Atlanta, Georgia.

"These organisms are very opportunistic," said Dr. Healey. "They enter a swimmer's body through the nose and go to the oxygen-rich environment of

7

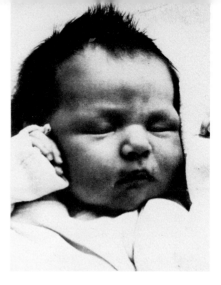

To stave off crib
death, a new German device
helps "remind"
babies to keep breathing.

the brain." There the amoebas devour brain matter and secrete a substance that kills the tissue. "We never knew amoebas could cause this kind of damage," he said.

Researchers are baffled by the erratic occurrence of the disease and its resistance to treatment. Most of the reported cases have involved young people, primarily boys. Only three victims have survived, one a nine-year-old California girl treated with antifungicide drugs. But the same treatment failed to save an eight-year-old South Carolina boy.

Dr. Healey said he doesn't want to scare anyone: "The disease is rare, and anyone worried about it should wear nose clips while swimming in lakes or ponds."

CRIB-DEATH CURE Most new parents are terrified by stories they have heard about presumably healthy babies who suddenly stop breathing during the night. The syndrome, called crib death, results when infant lungs simply "forget" to keep working. Now help is on the way in the form of a device that "reminds" babies to breathe.

According to inventor Marianne Schlaefke, of Ruhr University, in Bochum, West Germany, any infant can momentarily forget to breathe. When this happens, the fluid that surrounds the lungs becomes extremely acidic and in normal babies jogs a control mechanism in the central nervous system. The control mechanism tells the lungs to "restart," but in some babies this mechanism is defective, and instant death occurs.

Schlaefke's solution is to teach infants to react to changes in acidity *without* help from the control mechanism. Her new system, which relies on biofeedback techniques, consists of a fan and an instrument for injecting acid into the lungs of sleeping infants. Whenever the acid is administered, the fan (directed toward the baby's face) creates a slight breeze that stimulates breathing. The infant soon learns to associate acid with breathing, and after a while the device is no longer needed.

Over 20 percent of
heart attack victims in one study
had been in
an "intense emotional state."

The new instrument is being tested in the children's ward of a Munich hospital. According to Schlaefke, however, when the tests are complete, parents will be able to use the system at home.

EMOTION AND SUDDEN DEATH It is a popular notion that emotional stress can induce a heart attack. Until recently doctors have had to rely on post-mortem reports from distraught relatives when attempting to assess the role that emotions play in sudden death.

Now psychiatrist Peter Reich and a team of doctors at Brigham and Women's Hospital, in Boston, Massachusetts, have asked 117 survivors whether they experienced an "acute emotional disturbance" during the 24 hours preceding their heart attack. Twenty-five of the survivors (21 percent) had been in an "intense emotional state" before the onset of their attack. Most of them were in this condition for less than 1 hour, but some had been in a state of panic for up to 24 hours. These 25 patients had no previous heart trouble, Reich says, suggesting that it was emotional stress that triggered the attack.

Reich says that the heart attacks were prompted by any one of a number of things: public humiliation, a divorce, a death in the family, and even ordinary fights. One hospitalized patient experienced a life-threatening arrhythmia (heartbeat irregularity) whenever his wife left after visiting him: He feared she would be mugged on her way home.

FETAL KICKS During a normal pregnancy, a woman gets used to the frequent kicks and movements made by the fetus in her womb. The motion tells her that the baby is healthy and growing. But if movement in the womb stops or slows, an Israeli gynecologist says, the baby may be sending out a cry of distress.

According to Dr. Eliahu Sadovsky, of Hadassah Hospital, in Jerusalem, the average pregnant woman feels between 50 and 2,000 weak and strong kicks and rolling movements each week. But if a woman reports that move-

9

*"The abolishment of pain in surgery
is a chimera. Knife and pain
are two words in surgery that must
forever be associated."
—Dr. Alfred Velpeau, 1839*

ments have suddenly decreased to half their average rate within two or three days, her unborn baby may be critically ill or even on the verge of death. To diagnose the condition of the fetus, Dr. Sadovsky immediately performs lab tests and then, if there is any danger, quickly induces delivery.

Some fetal health problems stem from the complications of a mother's disease: diabetes, for instance, or hypertension. Other fetal problems are due to unpredictable problems, such as pressure on the umbilical cord.

COMPUTERIZED M.D. It's 2:00 A.M. in the intensive-care ward, and the patient is still under heavy anesthesia from the coronary bypass operation he's had. Suddenly the alarms begin to ring. The patient's heart is beating erratically, his lips are turning blue, and his blood pressure is dropping. The nurse, having just completed her training, has had little experience with this kind of emergency. She pages repeatedly for a doctor, but by the time he arrives the patient has suffered a heart attack.

This scene is all too familiar in the medical world. But a team of physicians and engineers at Rensselaer Polytechnic Institute (RPI), in Troy, New York, is working on a solution: a medical autopilot, a computer that not only monitors a patient's vital functions but also directly controls those functions with respirators and drugs.

"The vital signs of a patient under intensive care must be kept within very narrow ranges for survival," says anesthesiologist Rob Roy, who heads the RPI effort. "There are so many variables to keep track of, in fact, that many intensive-care units already rely on computers to analyze all the data coming in." Physicians use the numbers provided by the computer to decide what drugs to administer.

To close the loop between the computer and the patient, the RPI team is building a series of "computerized doctors," starting with one that monitors the heart. According to Roy, the team has already developed most of the computer programs and hardware for this initial project. Sensors inserted in

Close look at a hair
follicle: Hair analysis can
help doctors spot
potential heart attack victims.

the hearts of laboratory animals connect to computerized equipment that
monitors heart function, Roy explains. This equipment is hooked to a central
computer that collates the information and, if anything is amiss, instructs a
system of pumps and valves to administer the appropriate drugs.

This is just the beginning: A private firm has also contracted with the team
to develop a similar autopilot for monitoring human patients.

HAIR DIAGNOSIS In the not too distant future we may well be going to the
hairdresser not just for a shampoo and trim but for a cardiological check-up
too. Research by a Hungarian cardiologist, Dr. József Bacsó, shows why.

Hair is a very suitable material for analysis in man. It's always easy to find
some, even when the subject is bald. And its content of various minerals
reflects closely the uptake of these materials, as well as their metabolism in
the body.

Working at the Institute of Nuclear Research of the Hungarian Academy of
Sciences, in Debrecen, Dr. Bascó observed that the calcium content of hair
from patients who had suffered a myocardial infarction was about five times
less than in healthy people. And he found the same telltale sign in hair from
infarction patients in a coronary-care unit several months *before* the onset of
coronary occlusion. So measurement of hair-calcium level may be a simple
way to predict the likelihood of a future heart attack.

There are, of course, a number of risk factors that seem to play a role in
promoting coronary artery disease and myocardial infarction. They include
smoking, high blood pressure, and obesity. So it's interesting that Dr. Bacsó's group has also found a close inverse relationship between risk factors
and hair-calcium level: The higher the risk factors, the less calcium there
was in the hair.

These results suggest that reduced hair-calcium level may be of value in
diagnosing narrowing of coronary arteries and, thus, in suggesting preventive measures. More studies are needed to confirm the usefulness of such

*"Injuries caused by the cold
include all those
due to lack of warmth."*
*—International Civil Defense
Organization*

an unusual screening method. But in the long run these efforts are worthwhile: The fate of the patient could hang by a single hair.

BOWEL BULLETS Doctors believe that emotional stress is associated with "irritable bowel syndrome," a common malady characterized by alternating bouts of constipation and diarrhea. To record the gut's reaction to everyday strain, London researchers devised a small steel capsule that houses a pressure sensor and a radio transmitter.

Two capsules, each dangling from a thin string attached to the patient's teeth, are dropped down the patient's throat (patients are told to swallow the "pills"). Each bullet then hovers just above the small intestine.

By tuning into the intestine's distress signals, doctors hope to learn more about the brain-bowel connection.

The capsules monitor contractions in the small intestine round the clock. Information about pressure changes taking place is radioed by the capsules to a set of aerials, in the shape of a belt and wrapped around the patient's body. The radio signals are then passed along electronically to a miniature tape recorder.

The procedure is not as distasteful as it sounds. David Wingate and Roland Valori, developers of the capsule, note that sufferers of irritable bowel syndrome can tolerate the pills for two to three days.

During this period, patients are subjected to different types of stressful situations, such as driving in traffic. Patients who've bitten the bullets claim the discomfort is worth it, if the method reveals more about their ailment.

LIGHTNING RECOVERY Lightning kills up to 300 Americans a year. In the first few seconds after people are struck, their heartbeat and breathing stop, leading bystanders to assume that they're dead. Within three minutes they are.

But, according to meteorologist Dennis Thomson, of Pennsylvania State

Running pigs: A new study
shows that, if troubled by disease,
a jogging hog is better
off than a reclining swine.

University, in University Park, persons hit by lightning can be saved with cardiopulmonary resuscitation (CPR), a manual technique in which pressure applied to the chest restores circulation and mouth-to-mouth resuscitation maintains respiration.

CPR has long been used to treat heart attack victims; more recently doctors discovered that it is also valuable in reviving those briefly exposed to electric shock. The technique was extended to lightning victims, Thomson says, because, like electric shock victims, they suffer a disruption of the electric impulses running through the brain and the heart.

Thomson advises that CPR should be performed by a person trained in the technique until the lightning victim can be taken to a medical center.

JOGGING PIGS Are pigs that jog less likely to suffer heart attacks? Several years ago researchers at the University of California at San Diego studied *healthy* pigs and answered no. More recently they repeated the experiment by using pigs with heart trouble, and this time the answer was yes.

To start the recent series of experiments, Drs. Colin Bloor and Frank White placed a balloon around one of the three main arteries leading to each pig's heart. When expanded, the balloon squeezed the artery, restricting the flow of blood, and in a short time the pigs developed heart disease. Ten of the diseased pigs exercised daily by running on a treadmill; another ten lazed about.

In the active pigs, Bloor and White say, arteries leading to the heart enlarged and sent out new branches, bypassing the vessels damaged by the balloons. In the inactive pigs, however, crippled arteries changed only slightly, and the heart muscle deteriorated.

Conclusion: Exercise helps prevent heart attacks, especially when heart disease is already present, because exercise increases circulation to the heart. But be forewarned. If you have heart trouble, exercise only on your doctor's approval. And don't overdo it: Two pigs died of excessive jogging.

*"Science has conquered many
diseases, broken the genetic code,
and even placed human beings
on the moon, yet when a man of
eighty is left in a room with two
eighteen-year old cocktail waitresses,
nothing happens. Because the
real problems never change."*
    —Woody Allen, in Side Effects

SECRET MILK INGREDIENTS What's in milk? Water, protein, fats, sugars, vitamins, and salts—plus a dash of morphine and a pinch of bovine leukemia virus.

Dr. Jorge F. Ferrer and his colleagues at the University of Pennsylvania School of Veterinary Medicine have sounded a warning about the virus, detected in fresh milk taken from ordinary dairy cows. It infects more than 20 percent of dairy cows in the United States and causes leukemia in those animals that have a genetic predisposition to it. The virus also infects sheep and chimpanzees. In the laboratory, it can infect human cells.

Whether exposure to the virus in milk poses any health hazard is not known. There is no proof, but one recent survey indicates that in areas where there are a lot of infected cows there is a high incidence of acute lymphoid leukemia in humans.

Less troubling is the presence of a morphinelike substance in cow and human milk. According to scientists at the Wellcome Research laboratories, in North Carolina, people and cattle may ingest morphine while eating certain plants—lettuce for us, and hay for cows. What it does is anyone's guess. The amount is tiny, just a fraction of the dose given as a painkiller. One researcher theorizes that the morphine acts like the enkephalins (substances in brain and intestinal cells that serve as natural painkillers) and that this may help explain why so much milk is drunk by insomniacs.

CAFFEINE CONFUSION Heavy drinkers often sip a cup of coffee to sober up. But experiments carried out at Hull University, in England, suggest that coffee may have the opposite effect. Tests on eight volunteer drinkers showed that their mistakes doubled when caffeine was added to their intake.

According to Dr. Geoffrey Lowe, each volunteer drank the equivalent of four screwdrivers and then the equivalent of two cups of coffee. When they attempted a series of tests, pressing buttons in response to light signals, the speed of their reactions was significantly slower than that of control subjects

Empire State Building's microwave transmitters: For Samuel Yannon, the "safe" dosage meant blindness and early death.

who drank only the vodka. Dr. Lowe concedes his sample was small, but he plans to conduct further experiments with 96 persons. Asked why people believe that a cup of black coffee will help them sober up, Dr. Lowe says, "Alcohol depresses the brain, while caffeine stimulates it. So the obvious thing to say [is] that they cancel each other out. However, it appears that, together, they overload the brain and cause confusion."

MICROWAVE DEATH Samuel Yannon, a technician who tuned transmitters atop the Empire State Building, began to suffer the bizarre symptoms of microwave sickness in 1965. He lost weight and couldn't remember details; blinded by cataracts and prematurely senile, he finally died in 1974.

Not too long ago Yannon's widow was awarded the sum of $30,000 plus $57 a week for life from the New York Workmen's Compensation Board. The board declared that Yannon had died from 16 years of exposure to microwaves. This decision marked the first time a state court has admitted that such radiation can be lethal.

According to U.S. occupational safety standards, Yannon was working in a perfectly safe environment. The maximum acceptable microwave dosage is 10 milliwatts per centimeter, and he worked at a level of 1.5 milliwatts. But the standard hasn't been updated since 1966 and is actually based on levels set by the air force in 1955. The Soviet Union, by comparison, recently set occupational standards of microwave radiation at less than 1 milliwatt per centimeter.

Milton Zaret, an ophthalmologist who testified on Yannon's behalf, says his symptoms were similar to those of other people he has treated for microwave sickness. The early signs, which Western doctors tend to shrug off as simple stress, include insomnia, poor sexual performance, sweating, and anxiety. By the time cataracts, heart pains, and memory loss set in, the disease is no longer reversible. "It's repeated, chronic low-level exposure that does people in," Zaret says. "We don't know what 'safe' levels are yet."

The rabbit test
was replaced by urinalysis.
Now a blood pregnancy
test may prove to be even better.

BETTER PREGNANCY TEST Most women are still unaware of a new blood test that detects pregnancy and that monitors the health of an embryo within just eight days of conception.

Dr. Judith Vaitukaitis, of Boston University, originally developed the test to monitor tumors that secrete human chorionic gonadotropin (HCG), a hormone also produced during pregnancy. She soon learned that the presence of HCG was a more precise indicator of pregnancy than the universally used urine test. Indeed, the HCG test was 100 percent accurate, while urinalysis worked only after a missed menstrual period and even then provided no information about the health of the embryo.

Because the placenta produces less pregnancy hormone in instances when an embryo is unhealthy or poorly attached, Dr. Vaitukaitis explains, the new test is a crucial index for physicians who treat women who are in danger of having a miscarriage.

And it can be effective in the early detection of the life-threatening ectopic pregnancy, where the embryo develops outside the uterus.

VIDEO THERAPY Space Invaders, Breakout, and other video-screen Armageddons that lure fixated preteens to play parlors are now turning up in more sober places.

At the Veterans Medical Center, in Palo Alto, California, Pong or Air-Sea Battle or Breakout can help treat brain-damaged victims of strokes, accidents, or senile dementia, according to psychologist William J. Lynch, video therapy's putative father.

A thirty-nine-year-old accident victim, whose left-hemisphere damage caused slurred speech and flawed hand/eye coordination, recovered (in part) under the influence of Breakout. Learning to maneuver a bouncing dot to break through a video-displayed ''wall,'' the man progressed from 20 hits in April to 50 in May and June, when he checked out of the hospital and got a job in a distant state.

Besides keeping
kids off the streets, video
games are helping
to treat brain-damage victims.

Not everyone's progress is so dramatic. If an Alzheimer's disease (senile dementia) patient doesn't deteriorate, it is counted as a victory, Lynch notes. Games are matched to a particular patient's needs, and his progress is painstakingly charted. Pong, say, might be the answer for visual-field problems, while memory or verbal deficiencies might be treated with a word game like Hangman.

Though the results of video-game rehabilitation are difficult to assess—since more routine therapies are used concurrently—Lynch is sanguine about an Atari future. "It's probably more fun than anything else we do. For one thing, patients are motivated." He's now angling for microcomputers.

Learning-disabled children in Massapequa, New York, are also honing their skills to the tune of make-believe heroics and intergalactic warfare.

Psychologists Renee Okoye and Tony Hollander have reported "dramatic improvement"—measured by the Sensory Integration Test—in the motor coordination, hand/eye coordination, and spatial visualization of 25 children given a hefty Atari diet.

And the army is allegedly considering using Atari games as part of its modern tank training for GIs.

STONE AGE SCALPELS Surgeons may soon be using a Stone Age tool updated by modern techniques to perform delicate operations. Scalpels modeled on blades that the ancient Mayans crafted from obsidian (volcanic glass) not only are sharper, but they are also much less expensive to produce than steel or diamond blades.

Anthropologist Payson Sheets, of the University of Colorado, practicing what he calls "applied archaeology," is currently working with an eye surgeon to test the effectiveness of the obsidian knives.

"The fractured glass edge is vastly sharper than anything commercially available with a honed edge. A honed edge is limited in sharpness by the nature of the steel and the size of abrading particles used," Sheets says. The obsidian blades are roughly only ten silicon dioxide molecules in thickness (three billionths to four billionths of a centimeter).

"We have now gone quite a bit beyond prehistoric stone-tool technology," Sheets says. "We're combining the best qualities of the obsidian with modern techniques, such as cast molding bronze and improved glass composition, to get a tougher edge."

Though the blades are still experimental, Sheets believes it might be possible to make them commercially available for about $20. Diamond scalpels cost $800 and up.

The eye surgeon testing the blades found them, after what Sheets termed "fairly routine use," vastly superior to what he could get on the ophthalmic-instrument market.

Sheets, who in the early Seventies did graduate field work on ancient Mayan technology, says the use of obsidian blades has been traced as far back as 2000 B.C. They have been found from central Mexico to Guatemala and El Salvador. "So what we have here is almost four thousand years of research and development," he explains. Most of it was effectively eliminated by the Spanish conquerors, who disliked the ritual aspects of the manufacture of stone tools by the natives. They also wanted to make the natives dependent on steel so that they could trade for locally produced items. "It was a bad time to be an Indian," Sheets comments.

LISTENING TO BONES A runner stumbles and twists an ankle. His doctor takes an X ray but detects no injured bones. That afternoon, when the runner returns to the track, his ankle breaks. The perplexed physician's conclusion: The ankle probably had a hairline fracture in the first place, one so tiny that the X ray didn't pick it up.

There are, it seems, certain kinds of bone damage that X rays simply overlook. But a biomedical engineer at Rensselaer Polytechnic Institute, in Troy, New York, has come up with a solution: an extraordinarily precise

If this stirring view of the sea also stirs your stomach, nei-kuans straps may be of some help.

device that diagnoses bones by "listening" to high-frequency sound waves passed through them. According to the inventor, Hyo Sub Yoon, the new method is predicated on the fact that broken bones will absorb high-frequency sound waves far more readily than whole bones will.

The injured leg or finger, Yoon explains, is placed in a collar of high-frequency sound transmitters (or mini–hi-fi speakers) much as an arm is wrapped for a blood-pressure test. Then a receiver is taped to the body near the transmitters. The receiver monitors the waves as they pass from the transmitters, through the body, and into the bone suspected of being injured. Afterward the same amount is transmitted into a healthy bone. Finally the two tests are compared; the difference in absorption levels, if any, will correspond to the degree of bone damage.

According to him, Yoon's new method is superior to the X ray since it can detect minute injuries—microfractures, or tiny gaps—deep inside the bone tissue. "Unlike X rays, sound doesn't just scan or photograph the surface of a bone," Yoon says. "It penetrates the bone tissue and can even detect a single splintered fiber."

NAUSEA STRAPS Ancient Chinese acupuncturists relieved nausea by pressing tiny points on the inside of the wrists. Oncologist Daniel Choy, of Columbia University Medical School, has developed a pair of plastic straps that do the same thing: By applying pressure to those tiny wrist points (called nei-kuans by acupuncturists), the straps can alleviate nausea within one minute.

Dr. Choy conceived of the straps during a storm-tossed yacht race between Newport, Rhode Island, and Bermuda. When a wave washed the ship's pill supply overboard, he had to come up with a treatment for the seasick crew. He taught them how to press the nei-kuans with the thumb. It worked, but only if the pressure was continuously applied. Dr. Choy realized there had to be a better way. So he built the wrist straps.

For nuns in Sussex,
England, large plastic sacks of
dust are translated
into housing for elderly people.

Worn like watchbands, the elasticized straps exert constant pressure on the tendons inside each wrist. Dr. Choy believes that they inhibit nausea by triggering a direct electrical impulse to the brain.

The straps were tested on 100 seasick cruise passengers and, according to him, helped 85 of them. The straps also had a high success rate with postoperative and chemotherapy patients, pregnant women, and even hangover sufferers.

The new straps are better than Dramamine tablets, Dr. Choy maintains, because they produce no side effects and, unlike the pills, work even *after* the onset of nausea. With the patent pending on his invention, Dr. Choy says the wrist strap will soon be on the market.

DUSTY NUNS Household dust, collected by nuns, is being turned to good use in modern medicine.

The dust comes in envelopes, parcels, and car trunks to Holy Cross Priory, in Sussex, England, where the Sisters of Our Lady of Grace and Compassion gather it into large plastic sacks. These then go off in regular batches to Beecham Pharmaceuticals, a drug manufacturer based at Brentford, in Middlesex.

The eventual beneficiaries of this bizarre trade (Beecham pays 10 cents a pound for the dust) are people who suffer from allergic asthmas. Many asthmatics are hypersensitive to antigens present in house mites, which are found in astronomical numbers in household dust. Beecham's task is to use the purified dust to produce a vaccine—exactly as would be done with the measles virus or the diphtheria toxin.

All sides benefit. Allergy sufferers are relieved of their runny noses and sneezes. Beecham makes a nice profit from the vaccine it manufactures out of the ton of house dust collected each year. And the Sisters of Our Lady of Grace and Compassion have already used *their* profits to help run interdenominational housing for elderly people.

For one man, an illegible prescription misread by a druggist resulted in emotional problems—and breasts.

PRESCRIPTION ERRORS If a physician writes an illegible prescription, his patient may receive a deadly dose of the wrong medicine. For instance, one doctor's order for Ethatabs—a muscle relaxant—was misread as Estratabs, which is a female hormone. Before the mistake was discovered, the male patient took Estratabs for over a year and developed large breasts as well as psychological problems.

According to Neil Davis and Michael Cohen, of the Temple University School of Pharmacy, in Philadelphia, not all doctors write sloppily and medication errors are caused by more than just poor handwriting. The pharmacists, who have published a book titled *Medication Errors: Causes and Prevention,* say a breakdown in communication within hospital staffs is also at fault.

Abbreviations for instructions have sometimes caused the correct medicine to end up in the wrong place. The letters *O.D.* stand for the Latin *oculus dexter,* or right eye. But doctors occasionally use O.D. to mean "once daily." Nurses uncertain of the abbreviation have administered irritating oral liquid medications to the right eye.

Davis says that a patient can protect himself by using common sense: He should question the administration of eye drops, for example, if he is hospitalized for a condition unrelated to his eyes. To match drug and disease accurately, Davis and Cohen recommend that pharmacists maintain a patient's medication record, which would include a description of problems and past therapy. They also advocate that hospitals use the unit-dose system, in which the pharmacy provides the patient with daily medication in a labeled package.

RECTAL LASER Physicians in the United States generally use lasers in dramatic or exotic ways, such as to perform microsurgery or to reach inside the eyeball to treat blindness. But in China researchers are using lasers to cure such mundane ailments as toe pains and persistent anal itch.

Sharks are probably best known for their jaws, but in the future they may gain fame for their anti-cancer mechanism.

Doctors in the Chinese county of Zhao Qing were confronted with a high incidence of severe toe pains in young women. Some of the women were hospitalized because the pain was so severe that they couldn't walk. Three physicians at the local People's Hospital Number One, Chen Jin-he, Li Rui-xia, and He Ying-hui, turned to lasers to find a cure.

The painful toes were bathed in an invisible infrared laser beam for 15-minute periods. After four to six such laser treatments, the pain disappeared in all seven patients studied. The doctors think that heat from the laser beam cured the pain by expanding blood vessels and hence improving blood flow, speeding up local metabolism, and "possibly improving the mental state of the patients."

The same doctors used a different laser to combat an anal itch that had bothered one patient for 14 years. They illuminated the affected area with the red beam from a helium-neon laser for 15 minutes on each of 16 consecutive days, then repeated the series of treatments four more times. After a total of 20 hours of laser illumination, the itch "completely disappeared," the doctors report, but they cannot explain why.

SHARKS' CANCER SECRET Sharks rarely get tumors. In 26 years, after examining some 6,000 sharks, the Mote Marine Laboratory, in Sarasota, Florida, has found only one shark tumor. And it was benign.

If Carl Luer, of the Mote lab, is right, sharks may owe their tumor resistance to a natural mechanism that inhibits cancer. Once scientists learn how sharks avoid cancer, he says, they may be able to adapt the protective substance or system for the human body.

According to Luer, sharks perhaps stave off cancer with an immune system that, unlike our own highly specific system, destroys a broad range of invading substances. Their resistance might be due to the enormous quantities of vitamin A and fat contained in shark liver. There might even be an undiscovered agent, a chemical substance that confers cancer immunity.

*"An idea isn't responsible for
the people who believe in it."*
—Don Marquis

Luer plans to test resistance by feeding his sharks aflatoxin B1, a carcinogen produced by mold that is found on grain and ground nuts. Then, to identify the defense mechanism that keeps the sharks healthy, he will examine the chemical changes in their blood and liver. If and when Luer discovers the sharks' uncanny method of protection, human-cancer researchers will still have to apply the findings.

SPINE MEDICINE Two laboratory rats at Texas A&M's College of Medicine staggered a little when they walked, but walk they did—quite a feat when you consider that their spinal cords had been severed just a few months before.

The rats' comeback was due to an experimental treatment devised by Dr. John Gelderd. He placed the rats in a hyperbaric chamber. Over a period of two weeks he forced oxygen into the animals' tissues and administered DMSO (dimethyl sulfoxide). Six out of ten rats that underwent the combination treatment recovered some function and sensation, including the two that walked. Rats that received only oxygen improved too, but not to the same extent as the ones that received DMSO.

"When we sacrificed the animals, we saw new nerve fibers growing into the injury in the ones that showed return of function," Gelderd says. "We also saw the preservation of nerve tissue next to the injury."

The theory behind the new treatment, Gelderd says, is that a lack of oxygen in the area of a spinal cord injury not only causes intact nerve fibers to lose their function but also destroys nerve tissue. Supplying oxygen seems to encourage the growth of nerve fibers and to preserve tissue. DMSO is used because, among other things, it protects cells from the effects of injury. Since the destruction that follows a spinal cord injury begins almost at once, speed in treatment is important. Gelderd got his injured rats into the hyperbaric chamber within 20 minutes.

The oxygen-DMSO treatment obviously holds out hope for spinal cord–injured humans, but Gelderd cautions that his work, for the present, is strictly

Leeches have become serious medical tools once again, aiding blood flow and serving as temporary veins.

experimental. "This is basic research," he emphasizes. "It may take years before these methods are perfected."

BLOODSUCKERS A man lies in a hospital bed, leeches clinging to his fingers. A medieval medical tableau? Guess again.

It's a scene from the ultramodern world of microsurgery, where lowly blood suckers help maintain the blood flow in microsurgically reattached fingertips.

"Leeches do two things," says Dr. Jane A. Petro, a microsurgeon at New York City's Albert Einstein School of Medicine. "Tissue that has a good blood flow into it but not out of it will get swollen and congested. Leeches help thin the blood by injecting a natural anticoagulant.

"More important, they extract a certain amount of blood, decreasing the congestion—which might otherwise cause tissue death. In a sense we're using leeches as a temporary vein until new veins can form."

The leech technique, so far applied only to reattached fingertips, could theoretically be used "for any body part where you have problems with venous congestion," Dr. Petro notes.

Leeches are serving as medical helpmates in New York, Boston, and Houston and in France, too.

SECRET OF THE PYGMIES Homer and Aristotle regarded them as a fearsome people, engaging in savage battles with cranes and even attacking Heracles. Modern-day anthropologists describe them as shy, primitive folk, and brave hunters. But there is one trait all chroniclers agree upon: Pygmies are short.

Scientists from the University of Florida have discovered what keeps pygmies from growing to a "normal" height: a hormonal deficiency.

Doctors compared the blood of 11 Binga pygmies hailing from the Central African Republic with that of 31 average-height people.

*"Women are much fiercer than men.
Nobody has ever given us
weapons for very long, have they?"*
—Margaret Mead

Although the Africans had the usual levels of a body chemical known as human growth hormone, says Dr. Thomas J. Merimee, who headed the study, their blood contained a third as much insulinlike growth factor I (IGF-I), an obscure body chemical.

The lack of this chemical restricts a pygmie's height to between four feet and four feet eight inches.

Because scientists have pinpointed the source of the pygmies' small size, they may now be able to manufacture the substance and give it to others with IGF-I deficiency. Dr. Merimee, who is with the University of Florida's Department of Medicine, Division of Endocrinology and Metabolism, found that IGF-I levels are also insufficient in some ethnic groups now living in the United States.

Dr. Merimee believes that IGF-I may prove more effective in promoting growth than human growth hormone, which is occasionally administered to slow-growing children. "A substantial number of people become immune to human growth hormone after a while," he reports.

No plans have been made to give the substance to pygmies.

BINGE DISORDER A team of University of Minnesota psychiatrists has isolated some of the causes for a strange eating disorder called bulimia, in which a woman will spend hours gorging herself with junk food and later get it out of her system by vomiting, laxatives, or both.

Drs. Richard Pyle, James Mitchell, and Elke Eckert studied 34 bulimia patients for nearly two years and found that the disorder is restricted to white, middle-class, upwardly mobile women who are obsessed with eating and being thin.

Women who tended to go on eating binges from 15 minutes to, in one case, eight hours were depressed and already on a diet when the binges began. What seemed to trigger some of the junk-food orgies, says Dr. Pyle, director of outpatient psychiatry at the University of Minnesota Hospitals and

White, middle-class, upwardly mobile women are the best candidates for the strange eating disorder called bulimia.

Clinics, was some sort of separation—breaking up with a husband or boyfriend, for example.

The binges took over the patients' lives. One businesswoman went bankrupt trying to support a $100-a-day food habit. Some others stole food or became kleptomaniacs to support their daily binges.

Women tend to start doing it young, Dr. Pyle says, when they are about eighteen years old and can become addicted to a daily binge/vomiting habit. He found one woman who had been doing this for 27 years, every day.

Like those who suffer from anorexia nervosa, the "starvation disease," all the bulimia patients are obsessed with being thin. Unlike anorexics, however, women who suffer from bulimia look normal, not withered and gaunt.

Bulimia is a health threat. One woman had to get dentures after regurgitated stomach acid ate away her tooth enamel. And bulimia's hold is powerful. Dr. Eckert asked bulimia patients which they preferred: a life out of control with bulimia or one in control as a fat person. All chose bulimia.

THE OUCH ROOM If you, the rugged and sensible adult, fear going under the knife, how much more anxiety must a child feel?

That's why the University of Chicago's Wyler Children's Hospital encourages its subteen patients to play surgeon before undergoing their own operating-room ordeals. In Wyler's Ouch Room skilled counselors guide the terrified young patient through many steps of the operation he or she faces.

Lynn Ochs, the director of the South Side hospital's Doctor's Play Program, tries to make the program as realistic as possible.

In a typical Doctor's Play the child is outfitted in a surgeon's smock, green paper hair net, and operating mask. Ochs and her staff have the kids sit at a table. Then they explain the stethoscope, tell them what X rays do, and, if the children are older than four, give each one his or her own hypodermic needle. During the grandmotherly briefing, the children work at a scaled-down operating table with their "patient"—a large and floppy teddy bear.

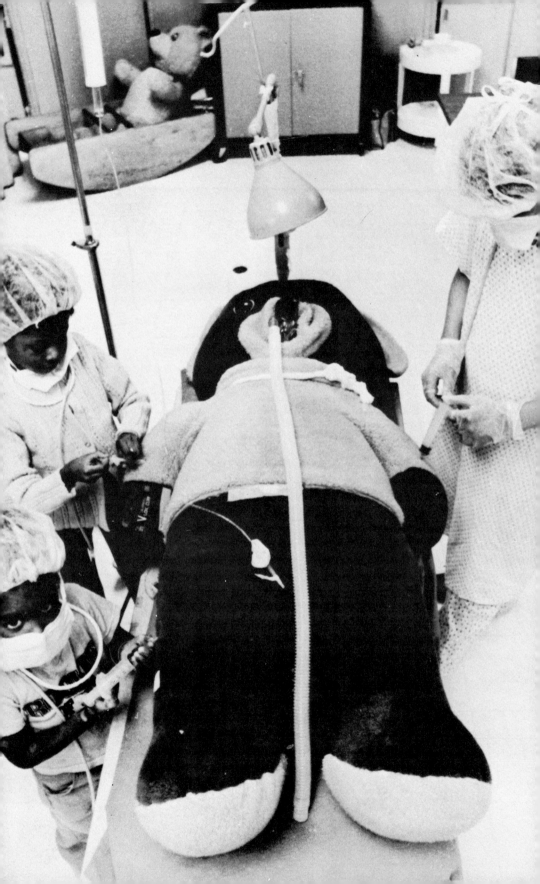

Kids in The Ouch
Room operate on their patient:
A very large
and floppy teddy bear.

The misconceptions children have about surgery can be appalling, Ochs says. One child was terrified that the surgeon would hack away at his "non-sick" areas. Others fear the hypodermic needle will leave a permanent hole or that they will never wake up from the anesthetic.

"We never, *never* use the word *scalpel*," Ochs adds.

SUPER SCAN An X ray or a CAT scan can find a tumor. But it takes a biopsy, an expedition under the skin, to determine whether that tumor is malignant. NMR imaging may soon change this diagnostic approach.

NMR (for nuclear magnetic resonance) imaging can identify malfunctioning chemical processes in the body before there are anatomical changes of the kind that show up on X rays. "It's like a window on the very early stage of medical pathology," says A. Everette James, Jr., chairman of the department of radiology and radiological sciences at Vanderbilt University School of Medicine, in Nashville, Tennessee. According to Dr. James, NMR imaging has the potential to diagnose cancer, heart disease, and stroke much earlier than X rays do.

The brain is the first place NMR imaging will be used, James says. It will look at brain metabolism, cerebral blood flow, tumors, atrophy, degenerative diseases, and even functions connected with thought processes.

One of the original developers of NMR-imaging technology, physicist William Moore, of Nottingham University, in England, describes how it works: "You put the subject in a magnetic field. Then you send in a radio signal and listen, and a signal with both a size and a characteristic decay time comes back out again."

To get cross-sectional slices at different angles, the CAT scanner uses motors to move the body or the X-ray apparatus. Since NMR works with a magnetic field, it can assemble any slice without the need for moving parts.

The researchers have been able to map out the meanings of the readings and the images produced. The aspect that most excites everybody involved

*"If you do not think about the future,
you cannot have one."*
—John Galsworthy

is the ability to look under the skin to assess blood chemistry, metabolic activity, and tissue composition without having to draw blood or use the knife. Even further down the developmental path, it may eventually be possible to use the accuracy of the magnetic field to destroy or modify unwanted tissues or chemical imbalances.

ICY OPERATIONS Cryosurgery, or surgery with an ice lance, has been around for over 20 years. And, thanks to better instruments, it's being used in an entirely new range of operations.

In many situations the deep-freeze treatment has advantages over knife surgery because there's less blood loss, little or no scarring, often no need for anesthesia, and a shorter period of convalescence. Skin cancers, precancerous changes of the cervix, hemorrhoids, prostatic disorders, and nasal polyps are treated routinely with cryosurgery because they can be approached directly with a frozen probe without any need for extensive knife surgery.

Probes used for cryosurgery are designed to emit universally available nitrous oxide or liquid nitrogen. Benign tissue is destroyed when subjected to temperatures of $-20\,°C$ to $-50\,°C$, while malignant tissue is destroyed at a minimum of $-50\,°C$. Cryodestruction works best when there's a rapid freeze at the lowest possible temperature. Where large tumors are involved, surgeons have been known to use the incredibly low temperature of $-180\,°C$ to destroy them.

During the past two years, Frigitronics of Connecticut, the world's largest manufacturer of cryosurgical instruments, has developed probes that can perform extremely delicate surgery. Neurosurgeons can now work with a 2.3-millimeter-diameter probe, containing a wire tip that is sensitive to precise temperatures while it is on tissue. Such a probe can freeze and lift tumors off delicate nerve centers without having surrounding tissue adhere to the probe's shank.

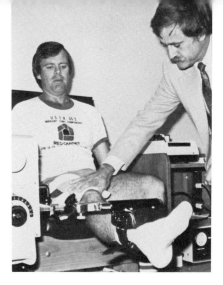

Now the runner's
"other" leg bone, the tibia
(or shin bone),
is getting some attention.

Ophthalmologists are using probes smaller than one millimeter in diameter to enter the eye and manipulate, position, and withdraw foreign objects and target tissue.

Dr. Bradley Rogers, of the University of Florida College of Medicine, in Gainesville, who has pioneered cryosurgery used to treat respiratory disorders in children, says, "These miniaturized probes open up new and exciting applications for cryosurgery in all fields."

NEW RUNNER'S PROBLEM Runners sometimes suffer what's called runner's fracture, a hairline break in the fibula, the smaller of the two leg bones. But doctors at Duke University Medical Center, in Durham, North Carolina, have found a new runner's fracture—in the other leg bone, the tibia (or shin bone).

Dr. Richard A. Daffner, associate professor of radiology, says the tibia break may be as common as runner's fracture but improperly diagnosed.

The break occurs a third of the way down the back of the shin, in a place where a muscle attaches to the bone. Dr. Daffner suspects it's a stress fracture caused by an improper or unnatural running gait.

Two of the first patients he and his colleagues saw were teen-agers who'd been diagnosed as having bone cancer. Dr. Daffner therefore urges doctors to get a complete medical history from runners who complain of suspicious shin pains and to "think fracture, not tumor."

He also recommends using radionuclide bone scans. "X rays may not always show a hairline break," he says, "but a positive radionuclide test says there is something wrong in the bone, even if the X rays don't."

Ballet dancers and gymnasts may also have this problem, but Dr. Daffner has seen it only in runners.

The cure for runner's shin break is a simple one, he says: Stop running for a month or so and let the bone heal. Otherwise, the hairline fracture could weaken and the shin bone could snap completely.

The video display terminal, or VDT, is becoming more and more common, and perhaps more and more dangerous.

VDT HAZARDS Video display terminals, or VDTs, those nifty computerized TV screens that seem to be everywhere—from newsrooms to airline reservation counters—can be a pain and possibly a danger to use, according to one government agency.

Since 1975 investigators from the National Institute for Occupational Safety and Health (NIOSH) have been checking out and largely substantiating complaints that heavy VDT use can cause eyestrain, headaches, enhanced stress, and backaches. And they are investigating yet another possible hazard: cataracts.

After a series of in-depth studies, NIOSH found that many of the VDT users' complaints were justified but could be eliminated with some design changes. Changing room lighting and using hooded and nonglare screens on the terminals would ease some of the visual stress, as would installing contrast controls and using soothing orange or amber lettering for the screens. Adjustable furniture and movable keyboards would help the muscle aches. Finally, NIOSH recommended that a 15-minute break from using the machines should come at least every two hours.

There is also worry that VDTs may cause cataracts, even though they give off radiation at levels considered safe by current government standards. Working at a VDT is equivalent to holding your face about two feet away from your television screen several hours each day, five days a week.

The first hint of trouble came in 1977, when two copy editors at *The New York Times* developed cataracts after using VDTs for about a year. After an investigation, the VDT was exonerated, but interest perked up again in 1980, when two other VDT users at another paper, the *Baltimore Sun,* developed the same eye problems. That has prompted NIOSH to do a comparison study of cataract cases in users and nonusers.

With about 3 million VDTs now estimated to be in use and more than 5 million expected by 1981, the need for accurately gauging the VDT risk factor will only increase as time goes on.

*"It is not all true that the
scientist goes after the
truth, it goes after him."
—Soren Kierkegaard*

RANDY YEASTS The sex lives of yeast cells, the microscopic molds used in making beer and bread, can give important clues to the causes of cancer. This is the view of two young scientists, Drs. David Beach and Paul Nurse of Sussex University, who have been studying the behavior of abnormal mutant yeasts, whose way of life involves nonstop sexual reproduction regardless of conditions around them.

"They're very confused cells that have sex when they really should be dividing. Because they're so sexy, we call them 'randy' mutants," says Dr. Beach. Normal yeasts go for sex only when conditions become unfavorable, because the exchange of genetic material involved in sex is followed by the making of tough, resistant spores. Any other time, normal yeasts reproduce asexually, by simply dividing in two.

Drs. Beach and Nurse have shown that the randy yeasts are unstable and that they are very frequently transformed—like reformed rakes—into yeasts exhibiting exactly the opposite behavior patterns. These reformed yeasts never go for sex at all; they just go on dividing asexually no matter how bad conditions get. The scientists, who have christened these "frigid yeasts," have demonstrated that the frequent transformation from randy into frigid yeasts is caused by a "jumping gene," a piece of genetic material moving from one position to another on one of the chromosomes in the yeast cells' nuclei.

The fact that a jumping gene can have such an effect may give important clues to the causes of cancer. Normal human cells stop dividing after about 50 cell divisions. But cancerous cells keep on dividing endlessly, like the frigid yeasts. Could this be due to a gene's, or many genes', making a jump to a wrong position on a human chromosome? It's an exciting possibility, and one that is made more likely by other research findings, which have revealed that genes jump around inside cells much more than had been thought. The idea of a stationary genetic blueprint is now out of date.

Other researchers have found very detailed similarities between viruses

The bane of athletes: the torn tendon, which can put them out of commission indefinitely. Now, a solution . . .

that cause cancer in animals and perhaps in humans and genes known to jump around in the cells of fruit flies. Jumping genes may control cell differentiation, and wrong jumps may prevent differentiation and lead to cell cancer. And cancer viruses may have evolved out of jumping genes. The true picture will be more complex than that, but at least it's beginning to emerge.

CARBON TENDONS A torn tendon can put an athlete out of commission indefinitely. Even when the tissue does heal, the area remains more vulnerable than it was before.

Now for the athlete who has everything: carbon tendons. Artificial tendons, which will sub for the real thing until the natural tissue can repair itself, may eventually be implanted in injured jocks.

It works like this: Doctors weave the "mock" tendon either into the remaining tissue or into the tissue they're encouraging to grow; the carbon frame, a sort of mesh supporter, takes over the natural tendon's functions.

Unlike other artificial body parts, which replace a defective or missing limb or organ, these artificial tendons actually help the torn tissue to heal itself. Once the real tissue mends, the prosthesis, which is composed of carbon fibers and polylactic acid (PLA), begins breaking down. The PLA changes to lactic acid and the carbon fibers degenerate into small particles that remain, harmlessly, in the general area of the new tendon.

Dr. Andrew Weiss, of Newark, New Jersey, developed the technique and has implanted his carbon tendons in 40 people so far. In his work with animals, Weiss found that it took new tendon tissue only about 63 days to overgrow its carbon scaffolding. In humans the process will take longer, but the result will be a tendon that's as strong and as useful as it was before the injury, Dr. Weiss contends.

Athletes aren't the only ones who'll benefit from the short-term replacement. Weiss has also tried the carbon substitutes in people whose tendons have been torn during auto accidents and in arthritis victims. For now he's the only one performing the surgery, although he plans to train other orthopedic surgeons in the technique, which is still in the developmental stages.

BLOOD PILLS In the near future, arthritis sufferers and other patients who must take medication on a regular basis may find their drugs encased in red blood cells, rather than in gelatin capsules or tablets.

Doctors at Aston University, Great Britain, and Geneva University are working on a technique in which whole, human red blood cells are dunked in the drug solution, then injected into the patient's vein. Once in the body, the cells release the medication slowly into the bloodstream.

The benefits of this method are many. First, it prevents wide swings in the concentration of the drug in the body, a drawback of oral agents. Second, it seems to reduce stomach and intestinal problems, sometimes side effects of drugs taken by mouth.

In addition, British researchers found that much smaller doses of arthritis medications were needed when the drug was packaged in a blood cell rather than administered by mouth. And because almost any drug can be sealed in the cells, the technique could be used in the treatment of a number of chronic diseases and ailments.

Doctors in the United States have used artificial red blood cells, coated with chemotherapy agents, to treat cancer patients. The injected cells then carry the drug to the site of the tumor, according to Thomas Drees, president of Alpha Therapeutics, the manufacturer of artificial blood in this country.

ALLERGY TREATMENT For years people allergic to certain foods had either to restrict themselves to carefully monitored diets or to suffer the effects of allergic reactions. These reactions usually come in the form of skin rashes, diarrhea, abdominal cramps, and asthma. Recently, though, two doctors working out of Yale's School of Medicine have found that one drug can turn some strict dieters into normal eaters.

Joggers who overdo it
can now be warned by a portable
computer that flashes,
"Slow down. Take your medicine."

The drug, crymolyn sodium, was first used to treat asthma, a disease characterized by repeated attacks of shortness of breath. But the Yale doctors, Samuel Kocochis and Joyce Gryboski, found that it could also help children unable to drink milk and eat soy proteins. Stomach-intestinal food allergies are similar to the reactions of asthma sufferers. Crymolyn sodium acts by stabilizing certain cells in these allergy systems and preventing them from releasing their potentially harmful substances into the blood.

The drug, however, is not designed for everyone with food allergies. "If you're allergic just to milk," says Dr. Gryboski, "you can stop drinking it and survive very well by drinking other things. But we felt that the main value of using crymolyn sodium was in people who had multiple-multiple allergies and were extremely limited in what they could eat, especially children."

Children allergic to such foods as beef, eggs, citrus fruit, cereal, and wheat, and who normally would have to maintain a strict diet, can now use this drug without worrying about adverse reactions. Although there are no major side effects to the chemical, "there is," adds Gryboski, "some hypersensitivity to the drug itself." Crymolyn sodium is approved by the FDA and is now marketable for the treatment of food allergies.

PERSONAL HEART COMPUTER A man is jogging when his heartbeat suddenly quickens; a portable computer strapped to his belt instantly beeps. Glancing down at the computer, the man sees a message flash in bright red: SLOW DOWN. TAKE YOUR MEDICATION. He follows the computer's instructions and in 20 minutes resumes his jog.

The man is a cardiac patient, and the computer, built at the Massachusetts Institute of Technology, could save his live. Developed by bioengineer Roger Mark, the three-pound computer listens to heartbeats through an electrocardiograph hooked to the patient's chest. When the computer detects a danger signal, such as an irregular rhythm, it beeps and then flashes instructions onto a lighted display that the patient can easily read and then obey.

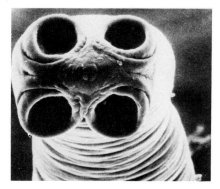

Tapeworms are responsible
for starving many people. But in
the United States
teen-agers do it to themselves.

"If the computer detects an especially fast heartbeat, it tells the patient to rest," says Mark. "If it detects signs of a heart attack, it instructs the patient to call a hospital." The computer records all irregular rhythms for physicians to examine later in the day, Mark adds. With this information, the doctor can gauge the patient's progress and prescribe treatment with more confidence.

TEEN MALNUTRITION Starvation diets are dangerous for everyone, but teen-agers who stop eating may suffer the most. Malnourished teens are vulnerable to permanent and devastating side effects ranging from stunted growth to infertility.

These conclusions come from the laboratory of Sanford Miller, a Massachusetts Institute of Technology biochemist who has spent the past few years starving rats. Miller found that rats were irreversibly damaged only when starved during specific stages of growth. "Each organ of the body has its own critical growth period," Miller explains. "If malnutrition interferes with these critical periods, permanent damage or deformities may result."

All animals, including man, go through critical growth periods during early infancy, says Miller. "But man is unique because he has a second growth spurt during adolescence. Thus, adolescents once again become susceptible to permanent damage resulting from malnutrition."

The rat experiments must be backed up by studies on humans before scientists understand exactly how starvation hurts teens, Miller notes. Meanwhile, teen-agers on fad diets might look out for such symptoms as antisocial behavior, inability to function under stress, dental problems and gum disease, and a greater susceptibility to illness, anemia, and food poisoning. A final word of caution: pre-teens from nine to twelve may also be vulnerable.

SHOCKING POISON Toxic shock syndrome has sent a wave of apprehension through many women in America in recent years. It seemed to strike

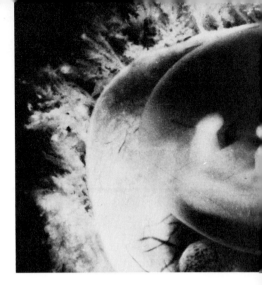

randomly, and its effects were capricious and, unfortunately, sometimes fatal.

The situation was made worse by the fact that researchers couldn't determine where the disease came from. It appeared to be related to tampon use, but the specific method of infection eluded detection.

Now researchers at the University of Wisconsin-Madison report that they've uncovered the cause of toxic shock. From early on, doctors suspected the staphylococcus bacterium of being the culprit, but they couldn't figure out how it caused the symptoms. The Wisconsin team tracked the bacterium to a previously unknown poisonous substance, called an enterotoxin, that is produced by some strains of the bacterium.

Enterotoxins attack intestinal linings and cause the vomiting and diarrhea associated with toxic shock. What made them so hard to detect in this case is that staphylococcus normally produces either poison or infection; with toxic shock it produces both.

JELLO LIVES If a bowl of Jello were wheeled into the intensive-care unit (ICU) of a modern hospital, would doctors there try to save its life with various machines and so-called heroic measures? The answer might possibly be yes, if physicians relied solely on sophisticated instruments.

In a now classic experiment, Dr. Adrian Upton, an Ontario, Canada, neurologist, hooked up a blob of lime Jello to an electroencephalograph (EEG) machine in an ICU. An EEG tells the doctor whether or not a patient has suffered "brain death." The verdict: The Jello was alive.

Dr. Upton explained that the tracings that appeared on the EEG resulted from influences—sweat, intravenous-fluid drops, respirators, and hospital technicians—always present in an intensive-care unit.

The odd experiment was reported by Dr. Robert S. Mendelsohn in *The People's Doctor,* his Chicago-based medical newsletter for consumers. Mendelsohn takes issue with the definition of death as stated by the so-called Harvard criteria, worked out by members of that university's faculty.

Even the fetus is not immune
to the dangers of nicotine. Mothers
who smoke risk giving
birth to young with abnormalities.

Mendelsohn says that in the old days (before 1960), doctors rarely had trouble figuring out when a patient was dead. And he questions whether high-priced technology is really superior "to the old-fashioned finger on the pulse and mirror in front of the mouth."

"Unless," writes Dr. Mendelsohn, "there is life after Jello."

SMOKING BABIES Cigarette smoking has been blamed for everything from lung disease to yellow teeth. Now you can add yet another item to that growing list. Smoking during pregnancy may destroy a woman's chances of having a healthy baby.

That's the sobering conclusion a group of doctors at Wayne State University's School of Medicine, in Michigan, came to when they completed experiments on the effects of nicotine, a component of cigarette smoke, on pregnant rats. What Dr. R. Hammer and his associates found was not encouraging to would-be mothers. The chances of having a miscarriage or of giving birth to young with developmental abnormalities were increased in rats given nicotine.

For an embryo to become a healthy infant, certain conditions must exist inside the womb. Both adequate supplies of oxygen and important nutrients have to be available. Any slack in the mother's blood of these necessary supplies around the time an embryo is about to attach itself to the uterine wall may terminate pregnancy.

Dr. Hammer and his associates discovered that the nicotine absorbed from the smoke affected the environment of the womb even before the embryo had a chance to attach to the uterine wall. Nicotine, they found, not only reduced the blood flow to the uterus at this critical time but also reduced the amount of oxygen the uterus receives. The result: spontaneous abortions and abnormal young.

Dr. Hammer admits that the doses of nicotine used in the rat experiments exceeded those encountered during cigarette smoking. But, he adds, "even

*"I think and think for months
and years. Ninety-nine times, the
conclusion is false. The
hundredth time I am right."*
—*Albert Einstein*

the small amount of nicotine inhaled during cigarette smoking is sufficient to evoke fetal [oxygen deficiency]."

BLOODLETTING FOR HEALTH Back in the first century, the physician Galen suggested that men should bleed themselves once a month, since menstruation seemed to have beneficial effects for women. It turns out that he might have been right. An article in the British journal *Lancet* suggests that bloodletting may help prevent heart disease.

The author, Jerome L. Sullivan, of the University of South Florida College of Medicine, states that men in our society suffer from severe heart problems, while women who are menstruating don't. The reason, he says, is that "increasing heart disease in American men is associated with progressive accumulation of stored iron." When ovulation ends, the percentage of stored iron increases in women, he suggests, and, thus, so does the danger of heart attack.

This is not due to hormonal changes or age, he feels. "The most parsimonious explanation is blood loss."

Sullivan's hypothesis is currently undergoing the acid test of experimentation—a comparison of regular blood donors with their nonbloodletting compatriots. If his notion is true, the donors should have lower rates of heart disease than non-donors.

In the meantime, Sullivan is taking his own advice. "I find the evidence very interesting and persuasive," he states. "I can tell you that since developing it, I have begun to donate blood regularly."

CANCER COMPUTER A series of algorithms—pathways by which computers handle the information put into them—developed for the National Aeronautics and Space Administration holds promise of making automatic diagnosis of early cancer possible.

The algorithms are designed to work with digitized images of body cells.

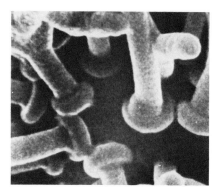

Nerves magnified 5,000 times. Scientists are studying the lowly leech for clues to man's nervous system.

The physician can note on the digitized image all the abnormalities he noticed in studying the patient's cells under a microscope. The computer then compares these features with characteristics stored in its memory and assigns an atypia status index number to the cell to indicate how abnormal it is. This number corresponds to one of five states ranging from simple metaplasia to definitive cancer.

In a preliminary test of bronchial cells from the mucus of cigarette smokers, the computer program assigned levels of cancer potential in keeping with the doctors' other information on each patient.

The program could be used in hospitals and laboratories to monitor smokers, industrial workers, and others with a high risk of cancer. It could catch cancer signs early, without surgery or the need for consultation with a specialist. The patient could have his cancer potential checked every time he gave a blood, mucus, or urine specimen.

LEECH NERVES Microsurgery has come very far, so that today severed hands, feet, and fingers can be reattached with nearly full use and nerve sensation. But doctors still know precious little about exactly how severed nerves reconnect and reestablish neural communication.

How, for instance, do the two halves of a severed nerve recognize each other, so that after microsurgery the attached appendage isn't a mass of crossed wires? Scientists at Cold Spring Harbor Laboratory have found an important clue to this process in, of all places, the nervous system of the slimy, inch-long North American leech.

Using monoclonal antibodies, the researchers stained specific cells and network paths in the leech's nervous system. The fact that one monoclonal could stain a whole network indicates that nerve paths may be marked by special chemical indicators.

While this doesn't fully explain how nerve cells make connections, it is a first step and opens paths for further research. For example, traditional the-

"Most people would die sooner than think; in fact, they do so."
—Bertrand Russell

ory has it that nerve information must be carried by surface proteins on nerve cells. But the monoclonals in this experiment bound to internal proteins, indicating that recognition may work from the cell's depths, not its surface.

The leech was chosen for these experiments because its nervous system has just 20,000 cells, the function of many of which is well understood.

EYE CALISTHENICS You might not expect an optometrist to ask you to read a sight chart as you bounce up and down on a trampoline, but that's only one of the surprises at Dr. Raymond Gottlieb's Eye Gym.

Other exercises the Santa Rosa, California–based Dr. Gottlieb asks his patients to perform include: blinking and breathing (taking two full breaths while rapidly blinking the eyes); palming (placing the palms over closed eyes); and reading letters from a spinning wheel while balancing on a teetering board. Dr. Gottlieb claims that these and other exercises improve vision by increasing mind-body awareness and reducing stress. They are designed to "let the body help the eyes coordinate their movement."

Combining two controversial vision-improvement methods, the William Bates exercises and Pepper Stress Reduction techniques, Dr. Gottlieb claims he has improved vision-related problems from nearsightedness to learning disabilities. His patients include preschool children and septuagenarians. "The majority of visual problems are stress related," Dr. Gottlieb said. "Our statistics show that eighty percent of the time people with bad vision can be helped without corrective lenses."

He includes himself among the successes. "Years ago I ran across a book on the Bates system and tried it," he said. "I was mildly nearsighted and wore glasses. After doing the Bates exercises for about a year, I didn't need the glasses anymore. You have to believe your own eyes."

PROTEIN SLICES In the study of many diseases, including cancer, physi cians have reached a point at which further progress rests on their ability to

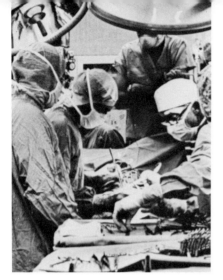

In delicate operations, computerized electrosurgery can prevent unnecessary cutting.

map and study the intricate relationships between the body's proteins, the worker molecules of our cells. Until recently, however, it has been extremely difficult to separate one protein from another and virtually impossible to get a complete reading of the state of all of a patient's proteins at one time.

At Argonne National Laboratory, the father-and-son team of Drs. Normal and Leigh Anderson has recently developed an isolating system that may make complete protein surveys possible. Called ISO-DALT, the process separates proteins first in thin gels by electrical charge, and then sorts each gel layer by molecular weight. It's like inducing a plate of tangled spaghetti to line up in rows so you can match lengths.

With this technique it is possible to compare and identify all of the 30,000 to 50,000 protein and protein fractions in the body. This will help doctors to determine whether certain proteins or protein levels are associated with certain diseases and to monitor the rate of genetic change in different kinds of people. If a certain protein pattern occurs throughout the lives of long-lived individuals, a key to healthy aging may be revealed.

Already the program has found patterns in blood and urine proteins from muscular dystrophy and several kinds of cancer.

COMPUTER SURGERY An automated setup now in the works may eventually enable surgeons to cut tissue and stop bleeding in delicate operations with more speed, precision, ease, and safety than ever.

The system, mainly consisting of a solid-state generator controlled by an 8080 microprocessor, is designed only for electrosurgery performed through an endoscope.

In electrosurgery rapidly vibrating blades, powered by high-frequency electrical currents, blast rapid-fire voltage in a fraction of a second to sever tissue or to cause blood to coagulate. An endoscope is a tube that can snake through the throat, and other body openings, and slide into a hard-to-reach spot to convey fiber-optic images back to surgeons watching a TV monitor.

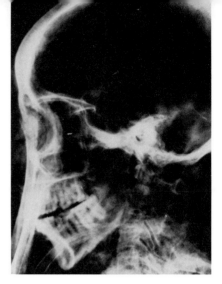

If the skull doesn't grow along with the brain, a child faces a life of severe headaches and mental retardation.

Surgical tools then slipped through a cable are manipulated like marionettes for the procedure.

Computerized electrosurgery done by endoscope—without actually opening up a patient—shows potential value against ulcers, polyps, other stomach disorders, and, possibly, such cancerous growths as melanomas.

The advantage here is that a surgeon can program the computer to control the intensity, frequency, duration, and rhythm of the electrical currents for each operation, preventing extra cutting or poor incisions.

"You can program an operation almost the way you compose music," says Otto H. Schmitt, biophysics-bioengineering professor at the University of Minnesota, who developed the computer surgeon with support from the National Institutes of Health and the Veterans Administration. "It makes the whole process faster and smoother because it has a program-it-yourself keyboard for the virtuoso operator."

After further experiments on dogs with ulcers, he says, the next step will be to test the device on humans within a year or two.

NEW SKULL The little boy, just seven, faced a life of searing headaches and mental retardation. The reason: a birth defect that caused his skull bones to fuse so that his brain had no room to grow.

But now, thanks to an operation at Children's Hospital Medical Center, in Boston, the boy has a new skull made of living tissue. The skull will grow to accommodate a developing brain, and the boy has a chance to lead a relatively normal life. The operation, pioneered by pediatric surgeon M. Judah Folkman, may correct even the most severe skeletal deformities. Dr. Folkman removed the top portion of the boy's skull and pulverized it to a paste. Then he molded the bone into the shape of a normal skull, using two pieces of the child's rib bone for scaffolding. The patient's body accepted the new tissue, which grew to healthy bone within a year. A similar technique may be used in the future for minor operations, such as correcting a cleft palate.

ANIMALS
AND WILDLIFE

CHAPTER 2

Porcupine perversions:
If you think hiding in pumpkins
is kinky, take a
look at this animal's sex habits.

ANIMAL LOVE "The variety of sex in the animal kingdom is amazing. Anything humans can imagine can be topped somewhere. And there's a lot they do we wouldn't dream of," says Dr. Robert Wallace, zoologist, behavioral ecology specialist, and a visiting professor at Florida State University.

Author of the book *How They Do It* (William Morrow and Company), Dr. Wallace cited some of the more bizarre animal mating habits. The male praying mantis, for example, may find itself being eaten alive. The female often pins her suitor to her thorax and begins devouring him, eyes first. She continues feasting, destroying a clump of nerves beneath his throat that controls his sexual behavior. Once these are gone, so are all inhibitions, and the headless insect frantically copulates, thrusting with greater abandon than was possible in life. The female saves the flesh that covers his sex organs for the last part of her meal.

Other animals are just plain cross, says Wallace. Porcupines take golden showers (the male urinates on the female before intercourse). A lonely female porcupine will stimulate herself with a stick.

Because humans tend to anthropomorphize (attribute human behavior to animals), they find the delicate manners of certain beasts interesting.

For example, one kind of turtle presses her legs together if she doesn't wish to mate; an unwilling female dolphin raps her suitor's head (or penis) with her flipper; the poison frog copulates only under a bed of leaves. The female empid fly won't satisfy her lover until the male performs a nuptial dance with a veil he has spun.

Homosexuality also flourishes in the animal kingdom. Bedbugs store the sperm of their male friends in their bodies and later deposit it in a female; geese form homosexual bonds, and if a jealous female interrupts the pair, a ménage à trois is born.

Although Wallace expressed surprise at the limited number of animal-human parallels, there is one trait common to both. Organ size isn't necessarily commensurate with the overall size of the animal. Though an aroused

> *"We must welcome the future
> remembering that soon it will be
> the past; we must respect the
> past remembering that once it was
> all that was humanly possible."*
> —*George Santayana*

whale sports a penis ten feet long and three feet in circumference, a full-grown male gorilla can boast a penis barely two inches in length. Snails have exceptionally long penises. And the penis of a dolphin is versatile; he uses this sensitive and flexible member to explore the ocean floor.

REFORMED KILLER BEES What do you get when you cross a tough African bee with a docile Brazilian one? A reformed killer. At least, that's what experts in South America are calling the once ferocious "killer bees" that made the headlines a few years ago.

The story begins like the plot of a B movie. In 1957 bees brought over from Africa to spark up the production rates of the lazier Brazilian bees escaped from their cages after a technician accidentally left the doors open. A few years later, reports of armies of "killer bees" stalking farmers filled the Brazilian news wires. Unlike other bees, which attack people and farm animals in gangs of 50 to 100, killer bees attack in swarms of 1,000. And even though the venom of the bees is the same as that of other species, a thousand injections of the stuff can be fatal.

Officials claimed that 16 deaths were due to these African murderers. The local tabloids were buzzing with rumors. One story reported how the bees had taken over the control tower at an Argentine airport. Another mentioned how the bees had wiped out the local bees in stinger-to-stinger combat. Actually, say scientists, the "killers" took over by moving into empty hives and establishing biological dominance in the area.

But the story seems to be ending happily. Since their escape, the Africans have been mating with the docile Brazilian bees and are now half as hostile as they were before. They have also become prodigious workers, according to the latest reports, and have helped to put the once-slackening South American honey industry back on its feet.

One note of caution, however: The reformed killer bees are heading for the United States. Bee experts expect them to pass through American cus-

*"As for man . . . he doesn't
even consider himself an animal—
which, considering the way he
considers them, is probably, all
things considered, the only
considerate thing about him."*
—Cleveland Amory

toms by 1985. Right now they're flying in Colombia. But soon the saga of the killer bees will be coming to a theater near you.

NUCLEAR GOPHERS The problem of where and how to bury the radioactive antitreasure from research facilities and commercial concerns has landed four pocket gophers jobs at Los Alamos National Laboratory, in New Mexico.

The gophers are getting their favorite foods—carrots, rolled barley, and a variety of green vegetables—for performing their normal burrowing activities. But since they are burrowing through seven feet of an experimentally arranged barrier containing topsoil, crushed volcanic rock, clay, stones, and gravel situated near a radioactive burial ground, the gophers should help answer an important question: How safe are low-level radioactive wastes in shallow-soil burial sites?

Gerald DePoorter, of the University of California, who designed the gopher experiment, plans to excavate the burrows to see whether the animals can penetrate waste-burial pits and carry contaminants to the surface. The project is part of a larger study, analyzing the interactions of delving plant roots, digging animals, and the erosive force of wind and rain with radioactive wastes. Findings from this work will be used to plan and build new burial sites where needed.

Not far from the gophers, 320 empty canisters of the type used to contain low-level wastes have been buried under fields of yellow clover, barley, and alfalfa, all of which have deep root systems. DePoorter and his colleagues will be trying various barrier materials to learn which ones most effectively detour the plants' roots around the canisters.

The gophers, DePoorter says, will be set free when their work is done.

MONKEY ORGASM Most scientists have assumed that female orgasm is unique to the human species, that females of other mammal species do not

Monkeys swinging:
Contrary to beliefs of zoologists,
our primate cousins
take sex beyond mere procreation.

experience the physiological changes that can accompany copulation.

Now an eight-year-old stump-tailed macaque monkey at the Netherlands Primate Center has proved them wrong. Telemetry signals from battery-powered transmitters implanted in her body showed strong uterine contractions and an accelerated heartbeat when she was engaged in sexual activity with another female. At the same time the monkey pursed her lips into an O shape, made rhythmic sounds, and manifested body tenseness—all signs of orgasm in male macaques.

This particular female was chosen for the experiment because she had shown what looked like orgasm when she mounted another female the way a male monkey would. Dr. David Goldfoot, of the Wisconsin Regional Primate Research Center, in Madison, and several Dutch colleagues implanted three transmitters into her body, two in the uterine area and one in the chest area. When she recovered from the operation, she engaged in the same sexual play with the same partner. The transmitters showed she was experiencing orgasm.

Her partner didn't show any signs of orgasm, but she did "kiss" the other monkey the way a female macaque often does when the male ejaculates during orgasm.

The first proof of female orgasm in monkeys came from female pairs, but Goldfoot has since obtained the same kind of telemetric evidence from heterosexual pairs. Simple visual observations of heterosexual couplings by macaques at the Wisconsin Regional Primate Center also suggest that some females had an orgasm, Dr. Goldfoot added. The frequency varies, some monkeys apparently never experiencing it and others experiencing it nearly half of the time.

"One of the reasons we are excited about these findings," Dr. Goldfoot said, "is that we now have a true animal model for female orgasm. We can ask questions about both the physiological and the psychological factors in orgasm."

Magnified photo of
an ant: In Brazil, these
insects are
served in a delectable sauce.

ASCENT OF MAN Jeremy Cherfas and John Gribbin, two British science writers, are monkeying around with a new theory of evolution. The monkey descended from man, they suggest, not vice versa.

Their argument is based on evidence that the DNA in both monkeys and humans was rather alike a mere 4.5 million years ago. This contradicts the fossil record, which suggests that a close kinship between man and monkey has not existed for 20 million years.

To account for the new genetic evidence, Cherfas and Gribbin hypothesized that approximately 4.5 million years ago a common ancestor of man, chimpanzee, and gorilla—a race of walking apes—split into two groups. One branch adapted itself to a rigorous life on the plains, eventually evolving into upright protohumans. The other branch preferred to hang out in the trees and eat fruit. They "de-evolved" into the monkeys and apes of today.

If this sounds like sheer monkey business, it is. "We don't really believe in the descending-ape theory," Cherfas says. "We simply wanted to show how many gray areas there are in fossil evidence. We'd like paleontologists to consult the molecular clock and then reconsider their findings."

So far their point has provoked no paleontological response. "We've heard from both preachers and politicians," Cherfas says, "but the fossil people have maintained a stony silence."

MICE WITH ANTLERS Mice with antlers may soon be providing important information to cancer researchers. Richmond Prehn, scientific director at the Institute for Medical Research, in San Jose, California, has succeeded in grafting deer-antler cells into mice and expects to be studying fully horned rodents in about a year.

"Antlers fall off every year and regrow very rapidly," Prehn says. "There's reason to think that tissue like that is relatively resistant to cancer formation." Prehn wants to learn why antlers can grow so fast, yet remain organized, while rapidly forming cancer cells will grow wildly.

Deer, however, would be too costly and inconvenient to study. So Prehn is trying to develop mice with antlers. "It's been found that you can transplant antlers from one site to another in deer. By transplanting a little bit of the surface of the bone underlying the antler to another site on the deer, antlers form where you put it," he says. He has taken samples of this tissue and transplanted them into rodents.

So far his tissue transplants have developed into small bumps, similar to the bases of deer antlers. But because of the huge differences between mice and deer "there's still some fiddling that'll have to be done," says Prehn.

TOASTED TERMITES Toasted termite tidbits, caterpillar crunchies, or french-fried flies? Whether or not we willingly choose such insect grub, most of us inadvertently eat scores of insect fragments daily. The FDA allows up to 20 drosophila-fly eggs in a glass of tomato juice, 75 insect pieces in a 2-ounce serving of cocoa, and frozen broccoli may have up to 60 aphids, thrips, or mites per serving. "It's impossible to eliminate all insects from food," says Cornell University entomologist Edgar Raffensperger, "but this represents no health hazard."

Entomophagy, or eating insects, may be by chance in the United States, but for many people it is by choice. Roasted termites are sought-after treats for many Africans; steamed giant waterbugs are prized in Laos, as are toasted stinkbugs in Mexico. Ants are served with a sauce in Brazil and with curry in Thailand, and in Indonesia crickets are seasoned and steamed in banana leaves. Commonly eaten insects around the world have included bees, caterpillars, cicadas (one of Aristotle's favorites), flies, grubs, lice, and even silkworms.

While most Americans cringe at the thought of eating such fare, we don't think twice about devouring an ordinary hot dog whose ingredients may include beef or hog scrotum, brains, lips, eyes, snouts, tailmeat, and stomachs. "Our aversion to certain foods is dictated by customs and habits,"

> *"The crisis of today is the
> joke of tomorrow."*
>
> —*H. G. Wells*

asserts Raffensperger. "Many insects are delicious and higher in protein, calories, and fat than equivalent amounts of beef."

In fact, protein and calorie content of flour and other processed foods could be doubled with insect fortification, resulting in no change in taste or appearance. Although the day Westerners dine on beetle bread is probably remote, insects could play a significant role in improving human nutrition in a world where millions of people are malnourished.

MYSTERY BIRD Luring a female bowerbird into a love nest is no easy task. The male of the species must first build an elaborate bower—sometimes up to eight feet high—from sticks, flowers, ferns, pebbles, fruit, and other colorful odds and ends. Then, using a twig or a leaf stem as a brush, he colors the bower with paints made from crushed fruits.

The wooing has just begun. The male sets a feast of brightly colored fruits, displays a morsel in his beak, and emits a strange love song. A smitten female will then join the male bird.

Just recently, Dr. Jared Diamond, of the University of California at Los Angeles, happened upon a yellow-fronted gardener bowerbird, a variety scientists had sought for decades but had never seen. During an expedition in New Guinea, Dr. Diamond, a physiologist and ornithologist, spotted a male standing before a four-foot-high nest, calling to a nearby female. As part of the ritual, the male raised his golden crest and caused it to quiver.

Dr. Diamond noted that the bird was unsuccessful in his lovemaking: The female, after watching from a branch, flew away.

Referring to the bird as the "Mystery Bird of New Guinea," Dr. Diamond said it's been the dream of many ornithologists to find this creature. Proof that the bird existed had been limited to three skins acquired by the American Museum of Natural History in the 1920s.

Unfortunately, most bird watchers will have to continue waiting for a glimpse of the yellow-fronted gardener bowerbird. Dr. Diamond's photos of

Helion and Robert:
The capuchin monkey fetches
food, turns the lights
off and on, and even vacuums.

the mystery bird and its courting scene were lost when his boat overturned and the film vanished.

FROG CRISIS In the tradition of the guinea pig, the rhesus monkey, Pavlov's dogs, and Skinner's pigeons, the frog has contributed to the progress of science as much as, or more than, any other laboratory animal has. It serves as a model for understanding heart function, endocrine function, and oxygen transport in higher animals, as well as for studying hemoglobin changes in humans, cloning, and Siamese twinning. It also provides millions of high-school students each year with what may be the most memorable event of their biology education: frog dissection. This is the view, anyway, of Richard J. Wassersug, a University of Chicago anatomist and frog expert.

All the while the frog was serving man, Wassersug recounts, man was bulldozing its ponds and migration paths to build highways, polluting the waters where frogs hibernate and lay eggs and where tadpoles grow to adulthood, stocking frog ponds with predatory fish, poisoning the frogs' natural food—the insect—with pesticides, and continuing to eat frogs.

Little wonder that this country now faces a growing frog shortage. The popular leopard, or grass, frog is disappearing from the northern United States, along with the bullfrog and the green frog.

More and more high-school students, Wassersug says, will now be dissecting fetal pigs instead.

MONKEYS FOR THE HANDICAPPED The blind have their dogs as aides and companions. Now the crippled may have monkeys to serve as their hands and legs. At Tufts-New England Medical Center Hospital, in Boston, capuchin monkeys are being tested as companions for patients who are confined to wheelchairs. The monkeys—they are the species used by organ grinders—will set tables, fetch books, turn pages, play records, open soda bottles, and perform dozens of other daily tasks that are frustrating or even

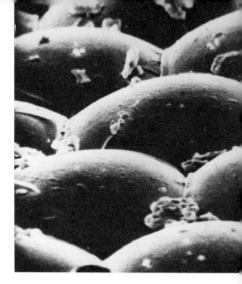

impossible for the severely handicapped. Capuchins, which come from South America, are quick, deft, and intelligent. Scientists think they can be trained to be both friends and servants to their human compatriots, who include more than 75,000 quadriplegics in the United States alone.

In one test, Helion, the monkey, shares quarters with Robert, a quadriplegic in his twenties. Robert can't use any of his limbs, but he can grip a small laser in his teeth. He uses it to point out objects he wants Helion to handle. He explains what he wants done verbally. Helion gets Robert food from the refrigerator, feeds him, vacuums the floor, and turns the lights on and off.

Dr. Mary Joan Willard, who directs the program, points out that a fully trained simian companion would cost just $6,000 to $9,000, in the same range as a Seeing Eye dog and a small price indeed for transforming the lives of the severely handicapped.

TRANSVESTITE FLIES It's been reported in hyenas and humans, mountain sheep and salamanders. Until recently, however, transvestism had not been known to exist among insects.

Female mimicry in male scorpion flies was happened upon by Randy Thornhill, insect-mating specialist at the University of New Mexico, during a six-year study of the insects.

Originally, Thornhill observed the *Hylobittacus apicalis* in hopes of understanding mate-selection criteria. The female, he noted, prefers males with large prey offerings. She keeps her genitals out of reach while assessing the nuptial gift. If the prey seems too small or unpalatable to the finicky female, she simply flies away before copulating or only copulates for a short while.

Females aren't the only ones eating and running, Thornhill found. A hungry male encountering another male displaying his booty turns transvestite, that is, mimics female behavior in order to get a fast meal. The imitator coyly drops his wings, wiggles his abdomen in the manner suggestive of a female, and draws back his genitals.

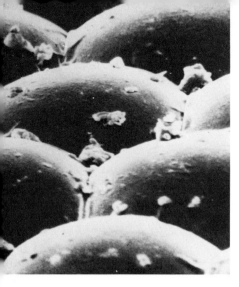

Close-up of fly's eye: One variety has discovered that female impersonation is one way of getting a free meal.

When the transvestite is successful in fooling his fellow fly and in absconding with the prey, he then darts off to find his own willing female. The next time he's hungry, though, he may not resort to such foul play. Although all males of this species possess the ability to exhibit transvestism (it's adaptive), the fly's behavior depends on the situation, Thornhill says.

Should the same male first see a tempting bug at dinner time, Thornhill explains, he'll feed on it. But, if the male first encounters another male with prey (or a copulating couple with prey), he'll attempt to pirate it.

The advantage of being transvestite? A longer, more productive life. Flies save so much time in food gathering that they can copulate more frequently. And they are exposed to fewer predators, such as the web-building spider, which often captures prey-hunting scorpion flies for its own dinner.

PETS AND BLOOD PRESSURE The next time you feel your blood pressure climbing, fondle the family dog. Studies at the University of Pennsylvania's School of Veterinary Medicine indicate that talking to (and petting) Fido can lower your systolic and diastolic pressure to below resting level.

Unlike humans, who may reject, refute, or contradict your conversation (and cause your blood pressure to rise), pets are totally accepting of their master's words and commands. Human-pet dialogues involve no risks, so the human relaxes, notes Dr. Aaron H. Katcher, associate professor of psychiatry at the Medical School of the University of Pennsylvania, who headed up the study.

Cats—even goldfish—have an equally soothing effect on people with hypertension and those with normal blood-pressure levels. Subjects instructed to "just watch the fish" repeatedly experienced a drop in blood pressure.

"There's been good evidence that looking into a fire, gazing at the ocean or at cloud patterns, interrupts normal thought patterns," says Dr. Katcher. "That's what's happening here. This just goes to illustrate a well-known general principle."

Chickens are in fact "bird brains"—and if they weren't the entire egg industry would be in big trouble.

Dr. Katcher's investigations did reveal some surprises. Of those surveyed during the five-month-long study, 78 percent claimed their animals can sense human moods and feelings. And nearly one third of the pet owners admitted confiding in their furry friends.

Researchers throughout the country have been so heartened by the therapeutic benefits of animals that pets are being placed with the autistic, the schizophrenic, and even prison inmates. Supporters of these therapies contend that animals can help evoke responses from the unresponsive and placate edgy criminals.

Indeed, man's best friend may make his owner's life longer and happier. In a study with heart-disease patients, Dr. Katcher found that 28 percent of the patients without pets died within a year of their hospitalization, while only 6 percent of those with pets died.

ANIMAL INCEST How to keep a baboon from mating with his mother or a favorite sister?

Inbreeding and the genetic defects that ensue are sometimes a problem for zoo administrators unaware of their animals' "roots." But an organization called the International Species Inventory System (ISIS) is helping zoos cut down on animal incest by providing vital statistics on more than 50,000 creatures in captivity.

Located on the grounds of the Minnesota Zoological Garden, the service keeps listings of the number of species in zoos, their whereabouts, the animals' age, sex, parentage, and other relevant information.

The 150 zoos throughout the world participating in ISIS (named for the Egyptian goddess of fertility) pay $1 per animal registered, at a minimum of 100 animals listed each year. ISIS then shares its data and inventory listing with members.

So if the Bronx Zoo needs a mate for its wolverine, the adminstrator consults ISIS to determine which sister zoo might supply one. Since the animal's ancestry is recorded, the zoo would then be assured of keeping the genetic lines clean.

Mating among siblings or cousins tends to lessen the chances of survival of the species. With wild populations dropping off, it's becoming increasingly important for captive animals to be self-sustaining.

"Certain zoo administrators used to think that if an animal died, they could just go out and buy another. Now they're becoming more conservation-minded," says Kim Hastings, administrative assistant for ISIS.

When ISIS first started in 1973, the service's main goal was to collect information on the blood chemistry of different beasts. This information is vital to zoo vets, who must treat dozens of different animals.

But this project soon took a back seat to the "animal mating game." Last year, ISIS began gathering blood data once again.

The animal inventory, along with physiological data, should help zoo administrators keep their animals thriving.

BIRD BRAINS Unlike our own behavior, which can change with the weather, birds tend to follow strict routines. Their behavior is encoded within their genes and, for the most part, is not learned. This can often lead to some rather interesting situations.

• Herring gulls lay speckled eggs. If they see a speckled egg outside of their nest they will roll it back in. The more spots an egg has, in fact, the more the gull will try to roll it back into the nest and sit on it. Shape doesn't matter. As long as the eggs are speckled, gulls will sit on square or even cone-shaped eggs and try to incubate them. Spots, it appears, are more important in egg recognition than either shape or size.

• Oyster catchers are impressed by large eggs. In fact, they will always choose a giant egg over their own. Even an egg as large as themselves will stimulate them to incubate it. Chicks hatched from large eggs, it seems, have a better chance of survival than those hatched from smaller eggs.

Are flamingos related
to whales? Zoologists have
discovered some
strange and striking similarities.

• When female pigeons are kept alone with other females, they don't lay eggs. But if the females happen to spot a male nearby trying to court them, they will lay eggs, even if their suitor is behind a glass wall.

• For most birds, egg laying stops after a certain number of eggs has accumulated in the nest. The house sparrow, for example, generally lays just four or five eggs. But if the eggs are removed as she lays them, the bird will continue to lay until as many as 50 eggs have been laid in succession and the ovaries are exhausted. Apparently, the sparrow needs to see a certain number of eggs in her nest in order to stop laying. Finding an egg missing, the bird will lay another to make up for the difference. The same behavior also occurs in chickens, which is the basis for the egg industry.

WHITHER THE FLAMINGO? A puzzling capability of flamingos has tantalized scientists since Charles Darwin first noticed it during the voyage of the HMS *Beagle.* The gangly birds, he noticed, could feed from South America's brine lakes, which were sometimes crusted with more than a foot of salt. Obviously, flamingos had some kind of impressive filtering system for separating salt and food.

Zoologists discovered, much to their surprise, that the flamingos' filter closely resembles that of the right whales. The whales suck water through a tight lattice of baleens, bones that jut downward from their upper jaws. Flamingos turn their heads upside down in the water to nibble on algae. The tongues of both animals, serve as a block for large chunks of food, and their lower jaws hold a filter through which small bits and water strain.

Can flamingos be related to whales? In a word, no. Studies by Storrs Olson and Alan Feduccia, of the Smithsonian Institution, indicate, rather, that flamingos are descended from the prehistoric bird Presyornis, a peculiar creature that combined the characteristics of a host of modern birds. The development of the remarkable similarity of the flamingo and whale filters, it seems, is a case of extraordinary evolutionary coincidence.

Flood-formed lake in
Australia: Home of the lungfish,
which can live without
food or water for five years.

WHALE LOVE SONGS Humpback whale songs have been an ''in'' science topic for several years, and a number of recording artists have played love songs against a whale-music background. Now research indicates that some of the whale sounds are actually love songs themselves.

Recent studies of whale songs have shown that they are incredibly varied and complex. They come in distinct styles, both by species and family group. And, it appears, they sing most often during specific seasons, especially during their journey to the mating ground.

The kind of singing humpbacks do during the mating trip is something like jazz. The whales in one group sing similar phrases, but never in unison. The songs are extremely long—up to 20 minutes—and are composed of phrases that are repeated several times, then followed by new phrases, all building to a theme that is repeated several times to make up the song. As the whales travel, they compose variations on the theme, building up one phrase, dropping another.

Zoologists feel these songs may be the male humpback's advertisement to a female mate. Like a Grandee crooning beneath a señorita's balcony window, these songs may convey the whale's emotional readiness to mate and convey vital information about him to help a female select the partner of her choice. Because the best, most complex songs have netted the best mates, whale singing has grown steadily more expressive and complicated. This interpretation raises the possibility that female whales may actually appreciate good musicianship. As animal behaviorist Peter Tyack wrote in *The Sciences:* ''Strangely enough, it may be the musical quality we sense in the males' remarkable songs that reflects the tastes of female humpbacks.''

SUSPENDED ANIMATION The lungfish may be the oldest existing form of life on earth, and one of the oddest. When the floods come to Australia's shallows, the lungfish lives an active existence. But when the waters recede, a strange thing happens. The fish burrows into the mud, leaving only a tiny

> *"The Buddha, the Godhead, resides*
> *quite as comfortably in the*
> *circuits of a digital computer or the*
> *gears of a cycle transmission*
> *as he does at the top of a mountain*
> *or in the petals of a flower."*
> —Robert Pirsig

hole for air. There it remains, in hibernation, without any intake of food or water for up to five years, until the waters return. Then it emerges unharmed and takes up life again.

Researchers at the Max Planck Institute, in Germany, have pursued the lungfish around the globe and have intensely studied its amazing ability to live on the edge of life without, apparently, aging or needing nourishment. Now they believe they have isolated and synthesized in the lab the chemical behind the process. And they think it might make the same state possible in humans.

The key chemical is a brain protein called a peptide that functions as a chemical information carrier. The Max Planck team has dubbed it amcurine. When laboratory-synthesized amcurine was given to test animals at the institute, their oxygen consumption and respiration rates were lowered and their body temperature decreased; all their metabolic functions slowed down. In cell-culture tests, cell division and DNA transcription were found to slow down when amcurine was present.

Most important, the tests showed that not only were amcurine's effects not toxic and fully reversible, but they were not species specific. Amcurine, a Max Planck report stated, "presumably functions in human beings."

Max Planck sees amcurine as having immediate applications as a tranquilizer, as a surgical aid to slow down life functions while difficult surgery is going on, and as a treatment for aging and cancer and other cell-related illnesses.

Underlying the entire project is the excitement of what amcurine's abilities really represent. The state induced by the peptide bears a startling resemblance to a favorite prediction of science fiction—suspended animation.

INSECT BRAIN TRANSPLANTS A dormant fruit fly wakes, expecting warm summer weather and plenty of ripening food—but instead is frozen in a winter blizzard. The result: insecticideless pest control.

Social lubricant:
If you want to meet people,
always take along a
dog—especially a pedigreed one.

Researchers at Ohio State University are using such methods as brain transplants among insects in an attempt to understand how their internal clocks work. "The implications," asserts entomologist David Denlinger, "are that by manipulating those internal clocks to make the insects wake early or sleep late, we may be able to control pest populations."

So far the scientists have isolated two insect hormones, both regulated by the brain, that seem to control their waking and sleeping cycles. The researchers study the way flesh flies and horn worms react when the brain of an insect in one cycle is placed in the head of an insect previously in another cycle.

The flesh fly, "the elephant of the fly world," weighs about 100 milligrams (it is about the size of a male thumbnail), and the horn worm is about the size of an index finger and thick as a thumb; both were chosen because they are large enough to work with easily.

"We're attempting to find out how the insect tells time," Denlinger said. "Our experiments show they measure nighttime very precisely. The insect's clock is located in its brain, and we're seeking the kind of chemical messages that might be involved. We want to know how this clock information transforms into a developmental signal.

"If found, these chemical messages might be artificially provided to confuse the insect. Since they are so finely tuned to the environment, the weather, a particular food plant, this could be an effective means of control."

Although it is possible that some insects might turn to an alternate food source, most "are fitted for only one food plant," Denlinger said. "For most, cold weather would be as effective as hitting them over the head with a fly swatter." Or spraying them with insecticide.

SOCIAL LUBRICANTS Everyone knows the dog is man's best friend, but now it looks as if our four-footed friend helps us with human contacts, too.

British scientist Peter R. Messent conducted a two-part study in which he

Could the Twist dance craze of the 1960s have been an imitation of bird courting rituals?

observed the effects of dogs on their masters' interactions with other people. In the first part of the study, dog walkers traversed an unfamiliar route in London both with and without their pet. People with dogs were contacted by other people 22 percent of the time, dogless walkers only 2 percent of the time. Not only that, but the dog walkers had a total of four short conversations while no one even said hello to the people without dogs.

One curious fact: People were friendlier in Hyde Park, one of the areas covered, but only to dog walkers.

Walking a dog in a familiar setting resulted in even more human interactions, Messent observed. Dog walkers taking their usual route talked to more people and for much longer periods than dog walkers in a strange neighborhood. The longest conversations were with fellow dog walkers. The dog walker—and the dog—had a little better chance of a human interaction if the dog was an obviously pedigreed animal.

What all this suggests to Messent, who is with Britain's Animal Studies Centre, is that dogs act as "social lubricants." The animals provide a means of getting people to socialize with each other more easily in both strange and familiar settings. It's a thought to bear in mind when the role of the dog in society is being debated, he adds.

TWISTING BIRDS Littering the trees of the South American forests are 50 different species of the vividly plumed manakin bird. What has ornithologists fascinated is not the color variety of the genus but the courting ritual practiced by one species, known as *Pipra filicauda*.

The courtship signal a male Pipra filicauda will give to his partner-to-be? A rather obscene twist of the tail.

An amorous male turns his backside to his lady friend, lifts his posterior, and begins stroking her chin with his tail feathers, which actually twist from side to side. A female that's been stroked before will know enough to position herself for the ceremony.

Polar bears: The animals are walking solar collectors, with fur composed of optical fibers.

The male's foreplay helps to forge a short-term friendship between the male and the female. As reported in the 1978–80 triannual *Report on the British Museum* (*Natural History*), this petting later becomes part of the ritual that precedes mating.

So far, this particular species of the manakin bird is the only one known to incorporate the twist into its lovemaking.

POLAR BEAR ENERGY Polar bears know something people don't know: how to collect the heat of the sun in a cold climate. If we follow the example of the polar bear, a Northeastern University professor believes, we may be able to improve the efficiency of solar collectors in colder areas of the United States.

Dr. Richard Grojean, an electrical engineer who serves as a consultant to the U.S. Army Research and Development Command, points out that the typical passive solar collector operates at less than 50 percent efficiency during cold weather. This happens not because there isn't enough sun but because most of the heat collected is lost to the cold air. The polar bear does a much better job.

"All the radiant energy that the polar bears do not reflect is transferred to their skin and used to warm the animal," says Dr. Grojean. "The efficiency at which they do this is phenomenal—up around ninety percent."

To find out how the big white bears do it, Dr. Grojean and his colleagues at Northeastern examined polar bear pelts. They discovered that although the hairs look white, they are actually transparent, with hollow cores. The internal scattering of light that hits the hairs makes them look white. The skin itself is black. Dr. Grojean and his colleagues theorize that sunlight, including the ultraviolet part of the spectrum, strikes the hairs and is transferred down them by means of radiation. The light is then absorbed by the skin.

"We think the hairs are really acting like the little optical fibers used in communication," says Dr. Grojean. He compares the hairs with the most

If you want a
friendly pussycat, choose one
that's been raised
in a family with lots of kids.

advanced kind of optical communication fibers. They have the ability to trap radiation and keep it from leaking out.

If the theory turns out to be correct, superefficient solar collectors containing an array of optical fibers might be constructed, Dr. Grojean speculates. One of his colleagues devised an experimental collector by putting fibers between the plates of a conventional collector. It increased efficiency by between 40 and 50 percent. Superior collectors based on fibers might also be used to produce electrical energy and to generate power.

FRIENDLY CATS If you want a friendly cat, advises psychologist Dr. Eileen B. Karsh, of Temple University, pick one that's been handled by humans from a very early age.

Dr. Karsh and her assistants have raised a number of kittens in the laboratory, some of which receive handling and some of which are simply exposed to humans. The handling starts around the seventeenth day. When the handled kittens become cats, they are almost invariably friendly. "Some of our experimental cats climb right up people's legs," says Dr. Karsh. "I can't wear stockings in the lab anymore." Nonhandled kittens, on the other hand, become rather aloof cats.

Dr. Karsh uses a number of measures other than ripped stockings to test friendliness. In one laboratory experiment, a person is put in a large box covered with wire mesh. A cat is put in a similar box. Then the test cat is brought into the room. Handled cats usually make for the box with the person in it, nonhandled cats for the box with the cat. In another experiment, a person holds the cat, making no attempt to restrain it. Handled cats stay in the experimenter's arms much longer than nonhandled cats. The champion, Petunia, will stay as long as she's held.

The amount of handling a kitten gets is apparently not important, Dr. Karsh says. One unusually friendly cat was handled just five minutes a day.

How can the prospective cat owner be sure of getting a friendly cat if he or

Wheat: Just a
nibble sends the female vole
into a flurry
of reproductive fervor.

she didn't raise the animal? You can't judge by a kitten's friendliness, according to Dr. Karsh, as most kittens are friendly up to the age of five months. She advises picking an animal raised in a family with lots of children. "Rough handling is better than no handling at all," she says.

WHEAT APHRODISIAC A female vole (a mouselike, mountain-roving creature) doesn't need flowers and springtime to stir her passion—just a nibble of winter wheat. Researchers have identified a substance in the new shoots of wheat and other plants that can send the most docile vole into reproductive fervor.

The aphrodisiac, synthesized by scientists at the University of Utah and tested on lab mice and voles, actually helps ensure that voles and other rodents common to harsh environs will continue to survive.

Found only in young, growing plants, the chemical cues the mammals to the fact that food supplies will now be available to their young and themselves and that they should start copulating.

Because the vole usually lives less than a year and has just one breeding season, timing is everything if the species is to survive.

The tundra-living lemming, known for its erratic population explosions, also relies on this food-source cue, report Norman Negus and Patricia Berger, two of the scientists who headed up the University of Utah studies. The stimulant may account for the sexual behavior of many mammals found in unpredictable environments, including the desert, where the time of food availability varies from year to year.

But sexual desire doesn't increase with larger doses of the chemical. Too much can even have a dampening effect on a vole's libido, researchers contend.

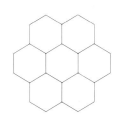

STARS
AND SPACE

CHAPTER 3

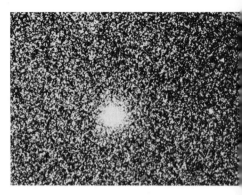

The search for
extraterrestrial intelligence has
focused on radio waves,
but viruses may be the answer.

VIRUSES FROM OUTER SPACE There may be no need to search for messages from space with radiotelescopes. Two Japanese biologists think interstellar "telegrams" may already be here waiting to be identified, in laboratories all over the world. They may even be inhabiting our own digestive tracts.

The proposed messenger is a phage, a virus that infects intestinal bacteria, called PhiX-174. When British scientists deciphered its complete gene structure two years ago, they found that parts of its DNA code could be read three different ways, depending on where the translation began.

The Tokyo researchers, Hiromitsu Yokoo, of Kyorin University, and Tairo Oshima, of the Mitsubishi-Kasei Institute of Life Sciences, think the code seems more artificial than natural. An advanced civilization, they suspect, may have applied to viruses a communications method already attempted by American radioastronomers.

The technique involves coding a message, either as radio pulses or as DNA bases, whose total number is a multiple of two prime numbers. The signals are laid out in rows like those in a television picture. A message 899 (29 × 31) signals long, for example, would be arranged as 29 rows of 31 signals or as 31 rows of 29 signals, to form a picture.

There are three possible messages in the PhiX-174 genes. One is 121 (11 × 11) units long, one is 91 (7 × 13) units long, and one is 533 (13 × 41) units long. Yokoo and Oshima have already tested the first two. "Unfortunately, nothing of significance has been found," they report. The scientists are currently working on the third possibility.

A message-by-virus system would have several advantages over radio, the biologists point out. There is no need for a receiver and an antenna aimed in the right direction at the right time. There needn't even be anyone waiting to receive a message. Landing in a suitable environment, the viruses would simply reproduce until intelligent life evolves to read the message. Which is precisely what Yokoo and Oshima are attempting to do.

*"All the signs suggest that life
exists on Mars, but
we can't find any bodies."*
—Gerald Soffen

SPACE GOODS There have been many spinoffs from the space program, from video games to life-saving medical equipment. But the biggest spinoffs are yet to come. Here is just a sample of the exotic space products the earth-orbiting space shuttle will provide in the future.

• Microgravity Research Associates Inc., in Miami, is preparing a machine for the 1983 shuttle flight that will grow gallium arsenide crystals. These will be purer than any made on earth and may someday replace silicon in electronic equipment. After more test flights, Microgravity hopes to send up a production-model machine to grow a commercial batch of crystals on shuttle missions every six months.

• John Vanderhoff, of Lehigh University, in Pennsylvania, will use a National Aeronautics and Space Administration (NASA) grant to make tiny plastic spheres in zero gravity on the third shuttle trip. There is already a market for the spheres. They are used, among other tasks, to measure the size of intestinal pores in people. Later on, though, scientists will make larger spheres in space, where the lack of gravity will prevent ingredients of different weights from separating into layers. If the process works, a company in Huntsville, Alabama, plans to market the spheres.

• One NASA scientist is working with TRW, Inc., on possible ways of casting turbine blades above the earth from single crystals of metal.

• Deere and Company is artifically creating a zero-gravity environment by using parabolic airplane flights over Huntsville, Alabama, in experiments with iron alloys. Later, it may conduct the same zero-gravity experiments within the shuttle.

• And an unidentified company in California is planning to fly metallurgical furnaces on the shuttle in order to sell zero-gravity wares to other shuttle customers.

A STAR NAMED SUE Now there is a way to become a star without having to audition or join an actors' union.

*"There is a coherent plan
in the universe, though I don't
know what it's a plan for."*
                    —Sir Fred Hoyle

The International Star Registry, a company based in Switzerland and with offices throughout the United States, is selling stars—the kind that twinkle in the night sky—to anyone who wants one. For $30 you can buy a sixth-magnitude sun anywhere in the known universe and christen it in your own name. For lovers, there's an assortment of binary stars—two stars that revolve around each other—for a mere $60. "Those are neat for anniversary presents and wedding gifts," adds a spokesperson for the registry.

What you get for your purchase is a parchment certificate proclaiming that your stellar name will live forever inside a vault under Geneva, Switzerland, and in a book within the Library of Congress. You also get a set of star charts telling you where you are in relation to all of the other stars.

So far, the idea has been a success. People are lining up to buy these stellar bodies. The company claims to have among its clients such celebrities as Barry Manilow, Mikhail Baryshnikov, and all of the cast from—what else—*Star Trek*.

But if the idea of owning a star appeals to many people, it isn't sitting well with established astronomers. Some call the company's idea illegitimate. Others call it downright stupid. "People shouldn't buy stars," complains one astronomer at the Hayden Planetarium in New York City. "All they get for their money is a certificate that they own a star. Big deal. Who's going to recognize that? And what good is it going to do them?"

The registry admits that astronomers won't be calling stars by their buyers' names in the near or distant future. "It's not really feasible for them to go out and find a star by these names," says a registry spokesperson. "They have to use some kind of coordinates." Today astronomers use nearly a dozen catalogs to plot their stars, and it isn't likely that they'll start to refer to the Star Registry.

But if you still want one, you can call the registry's toll-free number—800-323-0766—and charge a sun on your credit card. Considering the size of the company's stock, there's no hurry.

Martian surface:
A Russian study shows that
fungi, lichens, algae
and mosses would all survive.

DEATH ON MARS The question of life on Mars is far from settled. In June 1980 *Omni* reported that U.S. scientists had observed color changes in the Martian rocks around the Viking probe that might be due to biological activity. Now the Soviet Union has added another piece to the puzzle.

Working with data supplied by both the Soviet and American landers, exobiologists at the Space Biology Laboratory of Moscow University re-created a Martian environment on Earth.

They covered the floor of an airtight chamber with fine sand and lava, pumped out most of the atmosphere, and sent the rest swirling in artificial storms. Temperatures inside the chamber fluctuated between $+20°C$ and $-60°C$ while it was pelted from the ouside with X-ray and ultraviolet radiation.

Into this man-made Mars, the researchers introduced various forms of Earth life. Birds and mammals expired in a few seconds. Turtles survived for 6 hours, frogs and toads up to 25 hours. Several species of insects lasted for weeks. Oats, beans, and rye sprouted and grew, but none of them could reproduce.

Many of the lowest orders of life, however, adapted easily to the harsh conditions. Fungi, lichens, algae, and mosses developed and multiplied at a normal rate.

While this experiment doesn't prove that there is life on Mars, it does show that life—carbon-based life as we know it—is *possible* on Mars. And there are further implications.

Simply by selecting adaptable Earth organisms, we have the means to "terraform" our neighbor. In the future we might establish a primitive ecology on Mars that would fix carbon in the soil and release oxygen into the atmosphere. This, in turn, would enable us to introduce higher and higher life forms, eventually greening the Red Planet.

While such plans are no doubt ambitious, many scientists consider terraforming theoretically possible.

AMERICAN STONEHENGE For decades archaeologists have mined a massive Indian mound at Poverty Point in northeastern Louisiana. This incredible site consists of six concentric octagonal ridges with four broad pathways radiating from the center. The site is more than a mile across, 3,000 years old, and represents the high point of a mighty, unknown Indian nation that once traded all over the eastern United States.

But no one knew what, if anything, the location's odd configuration might mean. In autumn 1980, however, Boston University astrophysicist Kenneth Brecher visited Poverty Point and took careful measurements. A year later he announced that the mounds were an American Stonehenge, an ancient celestial computer. He found that one of its avenues pointed directly at the setting sun of June 21, the summer solstice; another aimed toward the setting sun on December 21, the winter solstice. Two other avenues point toward prominent stars that were used in the ancient world to keep time.

Although America's ancient astronomical observatory isn't quite as sophisticated as Stonehenge, it's much bigger. Fifteen million cubic yards of earth were moved to make it. This scale indicates that it was both a monument and the foundation for the Indians' dwellings.

SHRINKING SUN It may not look like it from where we stand, some 93 million miles away, but the sun is shrinking. That's the conclusion two scientists reached when they examined data recorded over four centuries of astronomical observation.

John Eddy, a visiting scientist at the Harvard-Smithsonian Center for Astrophysics, and Aram Boornazian, a mathematician, discovered that the solar diameter has been shrinking for the past 400 years. The researchers point to a solar eclipse that occurred in 1567. Observers in Rome saw a narrow doughnut of sun surround the Moon during the event. If the sun were the same size that it is today, the eclipse would have been total. Instead, it was not obscured by the Moon. That means, the two believe, that the sun

Stonehenge: Archaeologists
now say that America has its own
version made by
Indians in northeastern Louisiana.

was slightly larger 400 years ago than it is today.

After looking at the data and comparing it with other events, Eddy and Boornazian concluded that the sun has been shrinking at the rate of five feet per hour in the horizontal dimension and half that in the vertical.

Fortunately, the shrinkage doesn't apply to the entire mass of the sun, just to its outer layers. But since the speed of shrinkage is so high, the scientists think that the event is only temporary. It might imply, however, that this is part of a more lengthy solar cycle.

Astrophysicists are now trying to find out why this phenomenon occurs and how it fits into their model of the sun. Our sun, for example, gives off fewer neutrinos, subatomic particles thrown off by solar processes, than predicted in present-day solar theories. Eddy and Boornazian believe that gravitational contraction plays a role in controlling the sun's brightness. Meanwhile, the sun keeps shrinking. . . .

GOATS IN SPACE In outer space, man's best friend may be a goat.

According to computer simulations made by Cornell University researchers, working under a National Aeronautics and Space Administration grant, goats on interstellar voyages could gobble up wastes and provide astronauts with abundant food. The key, explains chemical engineer Michael Shuler, lies in the goat's versatile stomach—a chamber inhabited by microorganisms that break down all kinds of waste.

Astronauts can feed goats sludge, the woody parts of plants and other materials that people find unpalatable, Shuler says. The ravenous microorganisms present in the goat's stomach will break down these materials into feces and urine. The goat's feces and urine, ultimately fed into a mechanical treatment unit, will be far easier to process than the original refuse. The treatment unit will then convert the goat's excreta, as well as human feces and urine, into fertilizer for plants on the spacecraft. The plants in turn will produce food for goats and man, furthering the process.

Comet tail: Planets may also be so endowed, their visible portion just part of a messier celestial body.

With a goat aboard, Shuler says, the waste unit could be scaled down by more than half, greatly reducing the size of the spacecraft. This would save millions of dollars in fuel and other costs and would allow more of other types of equipment to be placed on board.

In return for this bounty, the goats will cost the human passengers virtually nothing, consuming only the wastes humans cannot eat. Moreover, they will provide milk and meat for the crew, lessening the amount of food needed on board. All in all, they may well make a ten-year sojourn in a spaceship far more feasible. The waste unit is soon to be constructed, but an actual flight may be ten or more years away.

PLANET TRAILS Back in the days of Einstein, before quantum physics clouded the picture, scientists thought the electrons surrounding the atom were discreet individual pieces, circling neatly around the atomic core. Now, it is believed they are more like comets: fuzzy, distended, and very hard to pinpoint.

A similar shift in thought seems to be occurring today in solar system astronomy. For generations astronomers have believed that the planets were sharply defined balls sailing through unclouded space. But evidence gleaned from space voyages indicates that that is not quite true. The sun and Jupiter unquestionably—and other planets most probably—appear to have long magnetic tails that wriggle off into space. Planets, it appears, may be merely the visible portion of a much larger, messier celestial body.

In August 1980, for example, *Voyager 2*, on its way to Saturn, passed through Jupiter's magnetic tail some two years after it had supposedly left the giant planet behind. *Voyager 2* was 390 million miles from the visible body of Jupiter when it passed through the tail.

Pioneer 10 has found that the atmosphere of the sun extends far out into the solar system, "far beyond the point predicted by many scientists," according to J. A. Van Allen, of the University of Iowa, discoverer of the magnetic Van Allen belts around Earth. The sun's atmosphere, it is now believed, extends 25 times as far as the distance from the star to the earth.

Instead of neat planets in clear space, the solar system may actually be fuzzy magnetic balls bouncing in thick solar soup.

SPACE-SURGERY SACK As large numbers of people begin living and working in Earth orbit for longer periods of time, eventually someone will need emergency surgery. It may be appendicitis, or a serious accident. And inevitably the day will come when a woman in orbit will need a delivery room fast—something the space shuttle *Columbia* doesn't have.

A West German researcher now has a solution. H. G. Mutke, of Munich, has come up with a light, transparent plastic sack in which are stored all the necessary tools for emergency surgery and child delivery in a weightless environment.

Various presterilized surgical utensils are kept inside pockets in the sack. Small magnets hold metal instruments down for easy retrieval.

In an emergency the plastic sack would be unrolled and fixed airtight around the patient by using an elastic band at the sack's rim. The mission specialists or astronauts performing the surgery, or the delivery, would inflate the sack by using a foot-powered pump and operate inside the sack by means of built-in, long-sleeved plastic gloves.

Once the operation is over and the body fluids (confined within the sack instead of floating around the ship) are sucked out, the sack can be removed, folded up, and stored for disposal.

Though the space-surgery sack hasn't been tested in orbit yet, it has been used in emergency situations in a Munich hospital.

HOT REAL ESTATE Looking for an out-of-this-world real estate buy? The Astronomical Society of the Pacific is offering deeds to land parcels on the planet Mercury in return for $25 donations.

Mercury's surface: Recreational sites of at least 14,000 acres are available for $25 donations.

The society promises "a nice recreational site, with a minimum of 14,000 acres," at that price, in a "pollution-free and also atmosphere-free environment." It warns prospective buyers, however, that temperatures are somewhat balmy—700°F (370°C)—hot enough to melt lead.

Grantees receive a deed, a high-resolution *Mariner* spacecraft photograph identifying the land area purchased, a summary of meteorological conditions on Mercury, and a table of information on the solar system.

The society is a worldwide, nonprofit scientific and educational organization that promotes numerous activities to increase public understanding of astronomy. It cites the disclaimer that it does not actually claim to own any land on Mercury. But then again neither does anyone else—yet.

COMET SHOCK The research satellite launched by the Defense Department in February 1979 was supposed to, among other things, study the sun's corona. But when a group of scientists working at the Naval Research Laboratory (NRL) analyzed photos taken from the craft, they found something quite unexpected: evidence of a comet smashing into the sun.

Astronomers do not usually get the chance to see a comet colliding with our sun. There have, though, been a couple of close calls. Seventeenth-century stargazers spotted perhaps a dozen comets passing extremely close to the sun. And there have been documented sightings of a solar-comet encounter as recently as 1945. Researcher Donald Michels and his NRL associates, however, believe that such an event took place on August 30, 1979.

The series of pictures taken from the satellite appears to tell the whole story. The first few shots show the head and tail of the comet as it approaches the sun. The next frames reveal the comet's head nuzzling up against the edge of the coronagraph's occulting disk, which is used to artificially block out the body of the sun. Other frames show only the tail. Finally, the comet disappears from view.

> *"Let your soul stand cool*
> *and composed before*
> *a million universes."*
> *—Walt Whitman*

As if these shots were not convincing enough, Michels points to later frames that show what appears to be evidence of a comet-splash into the sun: a halo of bright material surrounding the occulting disk.

Michels and his colleagues seem satisfied that a comet "death-plunge" actually did take place. But other scientists aren't so sure. They believe that the comet never crashed and that the halo was the result of the sun's illuminating the comet's tail material as it passed by. Yet none of the frames show the comet reappearing from behind the disk, as it should have if it had missed the sun. So the jury is still out, and only the comet knows for sure.

SPACE VAN If Len Cormier's plans work out, by 1987 we will have two choices of travel to space, the space shuttle and Cormier's compact alternative, which he calls the Space Van Orbiter.

Using mostly off-the-shelf components, Cormier believes it is possible to build a small fleet of long-nosed, stubby-winged space planes that can be piggybacked on a 747 high into the atmosphere. Somewhere above 40,000 feet, they would be released to rocket off into low Earth orbit.

Cormier, an aerospace engineer with 23 years of experience, has established Transpace, Inc., to design and launch that fleet.

The Space Vans will be small, about half the size of the shuttle, and consequently will carry smaller payloads—one metric ton as constrasted with the 29.5 metric tons the shuttle will hold.

Their big advantages over the shuttle will be in cost and convenience. At the peak of operation, Cormier says, the Space Vans could be making as many as 1,500 flights a year to low orbit, ferrying cargo and people for roughly one fifth of what the shuttle would cost. The Space Vans could be used for everything from disposing of nuclear waste to carting workers and materials to construction sites in space.

A lighter craft, the Space Van would heat up less on reentry and would not need as much thermal protection as the shuttle. Power will be provided by

*"We are such little men when
the stars come out."*
— *Hermann Hagedorn*

six RL-10 rocket engines, which, Cormier says, have been used routinely for space missions since 1962 and have been doing the job flawlessly.

$3 MILLION TOILET Astronauts have been complaining about it for years. Without gravity, they say, answering nature's call is a messy, totally distasteful chore.

That's why the National Aeronautics and Space Administration just spent $3 million on the new Waste Collector System, a high-tech space commode that simulates a gravity field so that it can be used like a toilet on Earth by both males and females aboard the space shuttle. The two-phase disposal system, which processes liquid and solid wastes separately, connects to a unit that resembles a regular toilet—except that crew members strap themselves onto the seat so they won't float away.

When an astronaut defecates, spinning blades beneath the seat create a powerful suction, which draws the waste down to where "it literally hits the fan," according to Frank DiSanto, the project manager for General Electric Space Systems, who refined the device over a three-year period. The rotating blades shred the waste and hurl it against the sides of a huge collector tank.

To "flush," astronauts pull a handle that slides a lid over the toilet, creating a vacuum that causes the waste material to vaporize through a filter and out into space.

Zero gravity poses a special problem during urination as well, since liquid released from the body does not flow in the expected stream. Instead, it separates into freefloating "spheroids." Astronauts find this circumstance disconcerting.

To get the stream flowing, engineers devised an air-suction tube that draws the liquid in through a funnel. Because people can use the toilet while standing or sitting, it works equally well for male or female astronauts. The liquid waste is stored in tanks in the belly of the spacecraft.

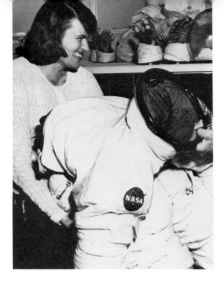

Astronaut Shannon Lucid suits up: Undergarments for zero gravity must also be specially designed.

According to DiSanto, the system passed muster on the space shuttle's shakedown cruise with flying colors, sparing the grateful crew the unsavory task of using and storing waste bags, which, he said, "were really a mess during previous spaceflights."

SPACE AGE BRA You wouldn't have to be an astronaut to benefit from a space age bra designed by a National Aeronautics and Space Administration engineer, the manufacturer contends.

Designed by NASA's Lawrence Kuznetz, the Support-Her bra employs the same theory of compression (rather than uplift) used in the design of astronauts' clothing. For spaceflight, it is important to ensure that bodies and clothing move in the same direction to prevent chafing.

Manufactured by Protogs, a Long Island firm, the bra combines spandex, velcro, carefully placed adjustable straps, and bottom ventilation to provide comfort to active women as they "shuttle around."

A product of an apparel-analysis course that NASA conducted at the University of Houston, the bra is only the first of numerous articles Kuznetz wants to design. Other ideas suggested include: garments containing packets of heat-absorbing gel strategically placed to keep athletes cool, children's clothing that grows as they do, and sleeping bags that can be converted into hiking suits.

BLACK HOLES RESCUED Black holes appear to have been saved from a theory that threatened them with nearly instantaneous oblivion. Still, physicists believe that black holes will eventually evaporate, and they aren't sure what will be left behind.

Although the gravitational field of a black hole is so strong that not even light can escape, subtle quantum effects do let matter trickle out. That leakage is so slow that it should take about 10^{73} (that's a 10 with 73 zeroes after it) years for a black hole as massive as the sun to evaporate completely,

A naked *what?* No, a
"naked singularity" is not obscene,
but simply what's left
when a black hole evaporates.

according to a theory proposed by Steven Hawking, of Cambridge University in England.

That theory was challenged in 1980 by Frank Tipler, of the University of Texas at Austin, who predicted that such a black hole would evaporate in only about a second. However, it looks like Hawking's time scale was right. "Tipler's equations weren't wrong," explains James Bardeen, of the University of Washington in Seattle. "He just misinterpreted the solutions." Although he wants to do more exhaustive calculations, Tipler himself concedes that "the evidence is against me."

Tipler had been trying to solve a problem that still remains with Hawking's theory: It predicts that evaporation of a black hole will leave behind a naked singularity, a point where the laws of physics break down. So far physicists haven't been able to carry more precise calculations all the way to the death of a black hole. Once a black hole shrinks to 10^{-5} gram or 10^{-33} centimeter, a quantized theory of gravity would be needed, Bardeen explains, but no such theory exists. Even if it did, it would require "horribly complicated" calculations, but physicists seem to find that prospect less unsettling than facing a naked singularity.

MOON BILLBOARDS A Los Angeles advertising firm wants to sell a new kind of space: billboards on the moon and minimessage plaques on the space shuttle *Columbia*.

The Bob Lorsch Company has taken these and other proposals to the National Aeronautics and Space Administration (NASA) and numerous government officials in what Lorsch calls an effort to "give the American people an opportunity to become involved" in the space program. Initially Lorsch wants to sell minimessage plaques bearing a "supportive, noncommercial" statement to 50 advertisers at $1 million each. The funds would be turned over to NASA, which would focus a camera on each plaque for 60 seconds during a shuttle mission and perhaps provide some VIP services to adver-

Physicist Paul Csonka
took a lesson from the spider
in designing a
cheap communications mirror.

tisers at a launch, according to Lorsch's scenario.

In seeking "exclusive outdoor advertising rights on the moon," Lorsch said, "it is inevitable that in the not too distant future there will be considerable amounts of traffic between Earth and the moon.

"Lunar mining and other development of the moon's resources . . . will in all probability be conducted by private consortiums working with government agencies. Advertising will be an ever-present, comforting reminder of the familiar things of their home planet," Lorsch says.

While it is easy to satirize the idea of moon billboards, Lorsch's proposals are aimed, he asserts, at generating revenue for the space program and producing public-relations materials to sell it. Several government officials and lawmakers, including Vice-President George Bush and Senator William Proxmire of Wisconsin, have expressed interest in Lorsch's proposals, although they returned the mock $50 million check made out to NASA that he sent.

POOR MAN'S COMMUNICATIONS SATELLITE Paul Csonka wants to put cobwebs in space, floating on beams of photons.

The University of Oregon physicist has received a patent (assigned to the now-defunct Department of Energy) for a cobweblike communications mirror that would float in near space above a specific spot on Earth. It would relay telecommunications signals from the ground to distant receivers, much as a communications satellite in geosynchronous orbit does.

The mirror would be a circular grid of closely spaced wires, not unlike a screen door. A balloon would carry it part way up; then the pressure of electromagnetic radiation beamed from a cluster of ground stations would push it the rest of the way.

Electromagnetic radiation consists of photons, particles that impart a force when they hit any object. Thus, the photon pressure would keep the dish aloft, hovering over one spot.

Have extraterrestrials hidden a space colony somewhere in the asteroid belt, some 200 million miles from Earth?

A cobweb mirror for shortwave communications would be about four meters in diameter, would weigh an incredibly slight 0.003 gram, and would hang suspended 100 kilometers above the United States from the force of a 150-kilowatt beam.

Such a mirror would cost about $100,000 to launch, Csonka says, and $130,000 a year to keep it up. Ground facilities would run about $450,000. By comparison, it takes $30 million to launch a communications satellite.

Csonka says his dish won't replace communications satellites, but it will have wide applications. Areas of dispersed population, such as the South Pacific, might find cobweb mirrors useful. So might underdeveloped countries that don't have the money to launch their own satellite or can't afford to lease the use of one.

Csonka's cobweb mirrors could well become the poor man's communications satellite.

ASTEROID NEIGHBORS Somewhere in our solar system's asteroid belt may be one colony or many of extraterrestrials—debating whether to join us or wipe us out, according to one astronomer.

Michael D. Papagiannis, chairman of Boston University's Department of Astronomy, says that if there is any other advanced, intelligent life in our galaxy, one place ideal for its space colonies is the asteroid belt, the band of orbiting rocks found between Jupiter and Mars.

The asteroids are excellent sources of raw materials for building. The entire belt is close enough to the sun to benefit from solar energy, and, Papagiannis notes, there are large gaps, called Kirkwood gaps, in which spaceships could park safely. Finally, a space colony, ten kilometers across or smaller, would be virtually impossible to spot from Earth.

Camouflaged by all this stone 200 million miles from Earth, the extraterrestrial colonists might well be studying us and keeping quiet as they observe our actions.

Science-fiction
squadron of spaceships patrols
wayward asteroids.
Experts now say we may need
such watchdogs to avoid disaster.

Why? "One can think of several answers," says Papagiannis, "including the zoo hypothesis, which is that they just want to see what we're up to. We've made tremendous technological progress in the last fifty years, building spacecraft and nuclear weapons. The extraterrestrials may just be hanging back, waiting to see what we will do next and debating whether to help us or, if we look dangerous, to destroy us."

ASTEROID BOMBS To learn more about asteroids and, not incidentally, to avoid disaster, scientists at the Jet Propulsion Laboratory, in Pasadena, California, and at the University of Arizona are working on a special plan to track these wayward space rocks.

Astronomers Tom Gehrels, at Arizona, and Eugene Shoemaker, of the Jet Propulsion Laboratory, have suggested as a project to the National Aeronautics and Space Administration a scheme that would include a device called a Spacewatch Camera, which would be mounted at the university's Kitt Peak Observatory.

With this specialized camera they could identify and plot the orbit of the asteroids, learn more about their physical makeup, give us some clues about the origin of our solar system, and even identify those that might be rich in minerals for space-mining enterprises.

As an added bonus, Gehrels notes, Spacewatch could save us from a global disaster. There are approximately 1,500 asteroids one kilometer and larger in diameter that have wandered from outer space into our solar system. If a large asteroid, say, ten kilometers across, ever hit our planet, it would have the equivalent impact of 10 million Hiroshima-size bombs.

Killer asteroids have hit the earth in the past. Just recently a group of scientists offered the theory that it was an asteroid that indirectly wiped out the dinosaurs 65 million years ago. The dust cloud raised by that collision killed off the plant life the dinosaurs ate. Some scientists also believe that it was an asteroid that caused the Tunguska explosion in Siberia in 1908.

With the Spacewatch Camera, Gehrels says, we could spot those aster-oids approaching too close to Earth and nudge them into a different orbit with explosives or rockets. The chances of a big one hitting us are small, about 1 in 100 million, Gehrels says. But given the tremendous amount of damage they could cause, it wouldn't hurt to be prepared. "We should be a little smarter than the dinosaurs were," he adds.

GRAVITATIONAL LENS Astronomers are seeing double, but there's noth-ing wrong with their telescopes. A distant galaxy is acting as a "gravitational lens," bending light from an even more distant object so that two images of the object appear close to each other in the sky.

Ordinarily light travels in straight lines, but its path can be bent by an in-tense gravitational field. Albert Einstein first predicted the effect in his gener-al theory of relativity, and it was first observed, by looking at light passing near the sun, during a solar eclipse in 1919.

The gravitational-lens effect showed up again when astronomers were surveying quasars—enigmatic objects that appear to be very bright and very far from the earth. In 1979 two quasars next to each other in the sky were found to emit identical light spectra. Such a juxtapositon is very unlike-ly, so astronomers began to suspect that the two images were produced when light from a single quasar was bent by the gravitational pull of a galaxy it passed on its way to the earth.

Some doubts surfaced when radio astronomers found that the radio im-ages did not match, but this, apparently, is due to imperfections in the gravi-tational lens. Astronomers at Palomar Observatory and at the University of Hawaii have found what appears to be the faint galaxy that's producing the dual image. Meanwhile, other astronomers have found a triple quasar image that also appears to be caused by a gravitational lens breaking up the light from a single quasar.

The observations may prove to be much more than a mere curiosity. The

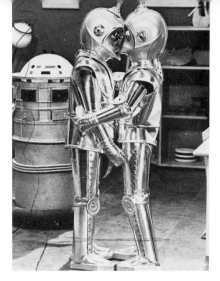

Robots may someday operate factories in outer space. They may also be able to replicate, though not via sexual reproduction.

two images fluctuate, but not at the same time because the light that makes them up travels different paths. By correlating these fluctuations, astronomers hope to measure the difference in the lengths of the two paths—and to get an indication of the radius of the universe.

SPACE ROBOTS In the center of an ash-gray crater on the moon stands a bustling factory manned by robots. Some of them scurry about on wheels, mining ore and transporting it to smelters. Others work in place, focusing television-camera eyes and flexing mechanical muscles as they assemble the newly smelted metal into spaceships, satellites, and, most important, new robots. Working in a vacuum under the lethal rays of the sun, the robots build and then operate additional factories that turn out more products.

Eventually, says the National Aeronautics and Space Administration, such robots could span the solar system. NASA has always used machines—satellites, spacecraft, landers—that might be called robots because they interact with the environment. But now, as scientists start planning the industrialization of space, they need far more sophisticated robots, truly intelligent machines that can see, hear, feel, smell, communicate, and move around—even make and carry out decisions.

NASA has recently decided to spend perhaps hundreds of millions of dollars yearly on robot research. In five years, says NASA, robots guided by their own brainpower might be able to fly from a space shuttle to a satellite. When it comes to the more complex task of repairing the satellite, a human supervisor will have to radio instructions to the robot brain; but in 10 or 15 years, space robots will be more or less autonomous, smart enough to build complicated structures or navigate rugged terrain.

Someday, according to Stan Sadin, NASA's deputy director of space systems technology development, robots with video ''eyes'' and propulsion systems will explore remote regions of the solar system. And self-replicating robots will build others like themselves, forming an infinite pool of labor.

Flight deck of the space shuttle: New lubricants are needed because in space no one can hear you squeak.

DRUGS IN SPACE The first space factory is on its way. The National Aeronautics and Space Administration (NASA) has joined with McDonnell Douglas Corporation and Johnson and Johnson in an agreement to send a pharmaceuticals factory into space in 1986.

The agreement is important not only because it represents the first business in space but because it is the first union of NASA and private companies for space development. NASA's role in the deal is like that of a developer who has set up an industrial park to encourage businesses to locate on his land. It guarantees transportation and adequate labor support (shuttle crew members will service the factory). Everything else comes from the businesses themselves, which makes NASA funds go much further.

The remaining big hurdle for the factory is power. It will require 3.5 kilowatts of electricity over several months at a time without recharge. NASA has several ideas for long-term power plants but isn't sure whether the money will be available to test any of them. In the meantime, shuttle flights will be carrying McDonnell Douglas space-drug-manufacturing equipment into orbit for tests. On the seventh shuttle flight in 1985, an 8,000-pound prototype factory will be taken up for a week-long test.

If all goes well, the factory will go aloft in 1986, and, if the FDA gives permission for marketing the space pharmaceuticals, Johnson and Johnson will begin selling them in 1987.

SPACE AGE LUBE You can squirt oil on a squeaky hinge at home, but in space traditional lubricants either freeze, burn up, or evaporate. The problems of lubrication in space have caused the creation of a new science—tribology, or the physics of friction.

The space age lubes must be able to withstand enormous temperatures. Oil works only up to about 375°F; synthetic petroleum products work to just 500°F. But some tribology products can withstand thousands of degrees of sustained heat without breaking up.

*"I could have gone on flying
through space forever."*
—Yuri Gagarin

The key to keeping parts moving in space is using solid materials with unique properties. Molybdenum disulfide, for example, slides against itself under pressure. So, it is excellent for such uses as protecting rotating antenna arms. Back in the old days, pioneers used to grease their wagon wheels with molybdenum disulfide dust when the pig fat ran out.

Other tribology creations are totally new high-tech products: supergraphites, soft metal films, and ceramic coatings that protect parts in even the most unimaginable situations. Literally hundreds of different space lubricants have been created for the thousands of parts that must be protected.

But tribologists are far from satisfied with their lubrication materials. The next step is the creation of composite building materials so that tomorrow's spacecraft have their lubricating ability built right into their frameworks. "Composites," says Donald Buckley, chief of tribology at NASA's Lewis Research Center, "are the future."

DIAMONDS IN OUTER SPACE When the *Voyager 2* robot spacecraft veers past Uranus and Neptune in the latter half of this decade, we may discover vast deposits of diamonds.

Up until recently, scientists have believed that the frozen depths of Uranus and Neptune were made mostly of methane. But the recent calculations of physicists Marvin Ross and Francis Ree, of Lawrence Livermore National Laboratory in California, suggest that high pressures and temperatures have separated the methane into carbon and hydrogen. Huge amounts of the carbon, they theorize, have been compressed into diamonds.

Since charged particles within the pressurized diamond layer would generate radio waves, says Ross, *Voyager 2* might detect the deposits during its trip through the solar system. Recovering this wealth, however, could be a long way off. Indeed, astronauts visiting one of these planets could amass untold fortunes in diamonds only if the surface—some 8,000 kilometers of gas—were to dissipate.

Once gravity waves are detected and understood, it may be possible to ride them as we now ride ocean waves.

GRAVITY WAVES FOUND AT LAST One of Albert Einstein's stranger predictions—the existence of gravity waves—finally appears to have been verified. Doing so required observation of an unusual pair of stars about 15,000 light-years from the earth.

Einstein's general theory of relativity, proposed in 1915, predicts that an accelerating mass should radiate energy in the form of gravity waves, much as electromagnetic energy is carried by light waves. However, gravity is much weaker than electromagnetic forces, and gravity waves were predicted to be so weak that Einstein himself wasn't sure they could be detected with certainty.

No one has yet directly detected a gravity wave in the laboratory. Physicists working in New England have had better luck in finding an effect of gravity waves—a slow shrinking of the distance between two stars orbiting each other.

Not just any two stars will do. What's needed is a pair of massive stars very close to each other and a precise way to time their orbits. Those needs were met in 1974, when astronomers discovered an unusual type of star called a pulsar, which orbits another star. Both stars are ultradense neutron stars, each packing 1.4 times the mass of the sun into a ball only about 15 miles in diameter. At their closest, they're separated only by the radius of the sun, while at their most distant, they're only five times the sun's radius from each other.

The rapidly spinning pulsar emits a narrow beam of radio waves that sweeps swiftly across the sky, producing a regular series of pulses at Earth-based radio telescopes. Working at the University of Massachusetts at Amherst, Joel M. Weisberg, Joseph H. Taylor, and Lee A. Fowler analyzed subtle variations in pulse timing and found that the orbit of the two stars is shrinking at about the rate predicted by general relativity theory. They say their results are "the first strong evidence for the existence of gravitational radiation."

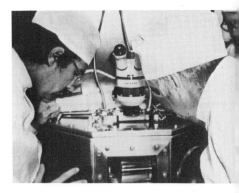

Scientists study
a pair of meteorites, called
shergottites,
that may have come from Mars.

MARTIAN METEORS A scientific team in Antarctica has found two objects that "may constitute one of the most exciting discoveries in recent years," according to Harry McSween, a University of Tennessee geologist who studied the find.

What engendered all the excitement? A pair of meteorites that scientists believe came from Mars. The two chunks—one the size of a lemon, the other the size of a cantaloupe—were found on a National Science Foundation expedition in Victoria Land. They belong to a rare family of meteors called the shergottites, named for the village of Shergotty, in India, where the first was discovered, in the 1800s. The only other discovery of a shergottite took place in Nigeria, in 1962. The Antartica find doubled the world's supply of the meteorites in one swoop.

Shergottites are unique in that they were formed by volcanoes and are incredibly young by meteor standards, just 1.2 billion years old, compared with 4.5 billion for most other celestial chunks.

Scientists speculate that the shergottites spewed from the Martian surface when it was smashed by some huge object. If true, the rocks represent a gold mine of information about Martian development. However, whether these rocks really come from Mars can't be proved until samples can be brought back from Mars for direct comparison.

BEHAVIOR AND THE MIND

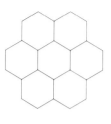

CHAPTER 4

Women's sexual ecstacy may outclass that of men, thanks to sensations bordering on altered states of consciousness.

HOTDOGGER OR HAMBURGER? When you go out for a quick lunch, do you prefer hot dogs or hamburgers? A Brooklyn psychiatrist recently completed a study that suggests the choice you make may reveal some aspects of your personality.

Dr. Leo Wollman conducted the study as a result of earlier research he did in 1976 for a book on obesity and dieting. "I got caught up in the subject of food preferences and did this study just for fun," Dr. Wollman said.

The study of 3,000 persons concludes that hot-dog eaters tend to be "outgoing, aggressive, ambitious extroverts," while hamburger fanciers are "quieter, introverted, more conservative types." Dr. Wollman describes hamburger eaters as a bit on the "wimpy" side.

"The people who eat hot dogs usually grab it and go," he said. "Hamburger eaters take more time. They're better-dressed executive types, used to making decisions—well-done, rare, ketchup or mustard . . ."

The sixty-six-year-old psychiatrist said there's an obvious "psychosexual" aspect to eating a hot dog: "The phallic symbolism, the way a person holds it, delicately or forcefully, the relationship to masturbation, and so on. But I didn't get into that much."

One fast-food chain lost interest in the study, and Dr. Wollman has decided not to publish the results, but he still believes it has "commercial value to advertisers."

WOMEN IN LOVE Having tasted both sides of life, the ancient sage/sex-changeling Tiresias (famous for his dire warnings to Oedipus) pronounced that sex was more fun as a woman. Two millennia later, brain researchers may be discovering why.

For men, sex is reflexive, a psychomotor activity, says neuropsychologist James W. Prescott, of the Institute of Humanistic Science, in West Bethesda, Maryland. Studies show that when the neurotransmitter dopamine, necessary to psychomotor activities, sinks to pathologically low levels (as in Par-

> *"I sometimes think that God in*
> *creating man somewhat*
> *overestimated His ability."*
> —Oscar Wilde

kinson's disease), sex is curtailed in men, but not in women. This leads some researchers to speculate that female sexual behavior is regulated by different neural pathways.

Psychomotor activity tends to be focused and goal oriented. Women's brains, less "focused" during sex, are better equipped to luxuriate in its emotional and spiritual dimensions, according to Prescott. So what our mothers told us was true, after all.

Female accounts of orgasm frequently describe sensations characteristic of altered states of consciousness—floating, loss of body awareness, a sense of unity with the partner. Prescott believes the vestibular-cerebellar system, a primitive part of the brain governing balance, touch, and movement, may account for these phenomena.

Female chauvinists, take note: According to Prescott, the human female is distinct from her mammalian sisters in experiencing sexual desire independently of estrus. Thus, for women, sex has a purpose beyond reproduction. Men's sexual makeup, conversely, represents no dramatic departure from those of lower mammals.

"All this is just speculation at this point," says a more cautious brain researcher, Jaak Panksepp, of Bowling Green State University, Ohio. "The only hard data are the studies with dopamine."

Panksepp's work with rats links dopamine with psychomotor stimulations. Rats with high dopamine levels ran around more, "self-stimulated" (pressed levers for rewards) more, ate more, responded more to the environment, and generally were more "outward-directed." Interestingly, women generally register lower dopamine levels than men.

WAR-NEUROSIS DRUG Traumatic war neurosis prevents an unknown number of veterans, particularly Vietnam veterans, from achieving peaceful, productive civilian lives. Veterans suffering from traumatic war neurosis have recurrent, frightening nightmares and vivid visualizations while awake,

Albert Einstein: The great genius's brain was found in a cardboard box under a beer cooler in Wichita, Kansas.

called "flashbacks," which morbidly repeat earlier traumatic events.

Generalized anxiety and panic attacks are frequently severe. Vietnam vets are often irritable and easily startled by sudden noises such as fireworks. Veterans can be explosive, violent, and aggressive.

Now thanks to recent experiments with the drug phenelzine sulfate, there is hope for these men.

Five patients with war neurosis had favorable responses to phenelzine. These were patients who had not responded to antipsychotics, tricyclic antidepressants, and psychotherapy with or without medication. With phenelzine, the patients felt calmer and stopped having nightmares and flashbacks. Startle reactions and aggressive, violent outbursts also stopped. Each patient felt much calmer almost immediately after treatment with phenelzine was begun.

Phenelzine sulfate is a powerful inhibitor of REM (Rapid Eye Movement) sleep: It completely abolishes dream activity in doses of 60 milligrams or more. All five patients stopped having traumatic nightmares while taking phenelzine; three who stopped taking the drug against advice began having the nightmares again.

According to Dr. George L. Hogben, who undertook the study along with his colleagues at the Bronx (New York) Veterans Administration Medical Center, the long-term outcome of traumatic war neurosis treated with phenelzine is uncertain. But so far, the veterans who continue to use phenelzine as advised are doing well.

Phenelzine sulfate may also prove effective in treating other similar disorders such as "atypical depression" or schizophrenia.

EINSTEIN'S BRAIN Whatever happened to the brain of the most celebrated genius of our time—Albert Einstein? Reporter Steven Levy of the *New Jersey Monthly* traced its whereabouts to the office of Thomas Harvey, medical supervisor of a biological laboratory in Wichita, Kansas. Einstein's brain is

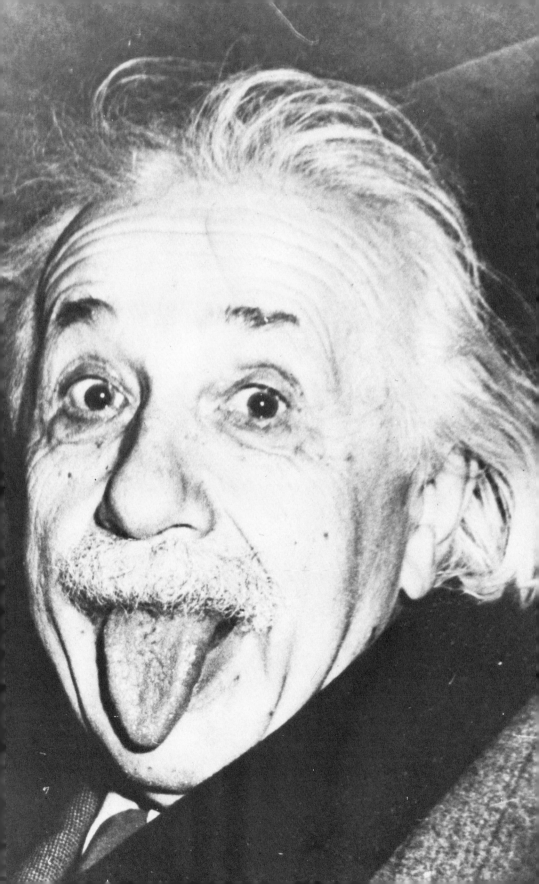

being stored in a jar of formaldehyde in a cardboard box marked Costa Cider under the beer cooler in Harvey's office.

When Harvey brought out the brain for Levy to see, the reporter rose to his feet only to sink back into his chair, speechless. In Levy's own words: "My eyes were fixed upon that jar as I tried to comprehend that those pieces of gunk bobbing up and down had caused a revolution in physics and quite possibly changed the course of civilization. There it was!"

Although most of the brain has been sectioned and given to various scientists for study, none of them have yet published their findings. All Harvey will say at this time is that the savant's brain falls "within normal limits for a man his age."

Many scientists feel that the physical examination of Einstein's brain will provide few clues to the secret of his genius. The brain of Karl Friedrich Gauss, for example, one of history's greatest mathematicians, when clinically examined, was found to be identical in weight, shape, and size to that of an ordinary day worker.

The key to spectacular intelligence does not seem to reside in dead tissue—the brains of morons have proved to be virtually indistinguishable from those of intellectual giants.

LUCID DREAMING With the right kind of mental training, it's possible not only to be an objective, clearheaded viewer of your own dreams but to step in and tinker with them as well.

So claims Dr. Stephen LaBerge, psychophysiologist and sleep researcher at the Sleep Research Center of the Stanford University School of Medicine, in Palo Alto, California. His claim revolves around what sleep researchers call lucid dreams, in which the sleeper knows he is in a dream while it is happening. "The lucid dreamer can reason clearly, remember frooly, and act volitionally upon reflection, all while continuing to dream vividly," says Dr. LaBerge.

The next time you
have a frightening dream, try to
make it "lucid" and
design a more pleasant outcome.

With this kind of objectivity, the lucid dreamer can step in and change whatever he doesn't like, the way a director or playwright instantly reworks the plot of a play.

Dr. LaBerge himself happens to be a natural lucid dreamer. As a child he managed to defuse a series of terrifying dreams about drowning simply by stepping in and giving himself the ability to breathe underwater.

"There is a talent for this," he explains, "just like everything else." But he says it is also possible to develop the knack with the right amount of motivation and a technique he calls MILD.

An acronym for the Mnemonic Induction of Lucid Dreams, MILD enchances the lucid dreaming of those who already do it and develops it in those who don't.

Very simply, it's a method of autosuggestion. As Dr. LaBerge gets ready for sleep, he says to himself, "The next time I'm dreaming, I want to remember I'm dreaming." Then he visualizes himself simultaneously lying asleep in bed and being in a dream and knowing full well where he is. With this simple method he managed to have an average of 21.5 lucid dreams each month and as many as 4 in one night.

Dr. LaBerge is interested in hearing from experienced lucid dreamers. He is at: Sleep Research Center, Stanford University School of Medicine, Stanford, California 94305.

SEPARATE CHECKS Most waiters and waitresses are reluctant to present large groups of diners with separate checks, but they would probably receive bigger tips if they did. "Tipping is cheaper by the bunch," reports psychologist Bib Latane.

Director of the Ohio State University behaviorial research lab, Latane originated the theory of "social impact," sometimes called "social physics" because of its quantitative nature. Essentially, the theory states that the way groups of people affect the behavior of individuals can be stated as a "multi-

Despite our liberated society, a study reports that women still tend to have more "submissive" sexual fantasies.

plicative function" of the number of people in the group, their status, and the immediacy of their presence.

In other words, larger groups have more effect on an individual than smaller ones, and high-status groups have more effect than low-status ones.

Experiments Latane conducted to test his theory produced elegant curves verifying his ideas—and some interesting specific findings. "Responsibility for individual action is diffused in large groups," Latane discovered.

Not only will people tip less in a restaurant when in a group, they also expend less effort in performing a task and are less likely to help someone in an emergency, he said.

"The sound of twelve hands clapping is not as loud for each individual as it is when he claps alone," Latane found in one experiment. In others, people ignored a knock on a classroom door when in groups and were less likely to aid others in need than when alone.

Latane said his findings suggest that "people work harder alone than they do in groups and may explain why groups of bystanders can watch and ignore someone in trouble. In the Kitty Genovese murder in New York," he said, "more than thirty people watched the victim being stabbed. Each of them probably saw the others and thought, Someone else will do something."

SEXUAL FANTASIES A new study on sexual fantasies goes a long way toward disproving popular psychoanalytic theory, which holds that fantasies are merely substitutes for the erotically deprived.

Researching the sexual fantasies of 901 students at Hofstra University, psychologist John Tantillo found that the heaviest fantasizers were also the most sexually active students—thus disproving the notion of fantasy as substitute. (Masturbation was considered a form of sexual activity.)

Incidentally, what Mother told you is true: Men spend considerably more time musing about sex than women do. "Despite the women's movement

Women with large breasts
were judged by both men and
women to be less moral,
less modest, and less intelligent.

and changes in sexual attitudes," notes Tantillo, "males are still reinforced to talk and think about sex."

Tantillo divided the groups' erotic themes into several categories. Men, he found, had more "general-typical" fantasies—such as oral sex, standard intercourse—than did women. Fetishism—"having sex with a life-size doll, being aroused by urination, enemas, mutilated bodies, and aprons"—also populated male daydreams, as did images of group sex.

Women, in contrast, reported more "submissive" fantasies, including rape and being tied up by a partner.

Most religions teach that "impure thoughts" lead to "sin." This precept may be more accurate psychologically than the Freudian theory that fantasy dissipates pent-up sexual energy, says Tantillo.

Another finding: While popular sex research by Shere Hite and Nancy Friday purveys the impression that everybody else is wild, kinky, and experienced, Tantillo notes that the typical college student in his sample has sex only once every two months or so.

MAMMARY MADNESS A study by a psychologist at Wheaton College, in Norton, Massachusetts, shows that big-breasted women are generally adjudged to be less competent, less intelligent, less moral, and less modest than women with small busts. This opinion was shared by both women and men polled by the researchers.

The study (for which we sincerely hope no large endowments were granted) included photographs of three naturally small-breasted female models who were asked to "stuff their busts," as Dr. Chris Kleinke, who headed the research project, put it, and then "stuff them larger." The photographs were shown to three groups, each of which contained 50 male and 50 female respondents, who were asked to rate the women on three aspects of personality based on first impressions.

Breast size had no effect on personal appeal (subjects were asked to rate

The games married
couples play around the time
of their wedding
is a good indication of the future.

the models with a series of adjectives, like "warm," "cold," "friendly," and so on), nor did it have any effect on whether the respondents thought the models bold or dominant. It was only the third aspect of the test in which male and female respondents alike concluded the best-endowed women to be immoral incompetents.

MATING GAME To Robert Ravich, psychiatrist and family therapist, how well your marriage works may be a matter not only of whether you win or lose but of how you play *his* game.

He has married couples try their hand at the Ravich Interpersonal Game/ Test, which requires each person to guide his or her own personal toy train from a start to a finish line in 30 seconds, more or less. The complication is that both persons have to do this act at the same time, trying to get to the finish line over the same two-track system. What is most revealing to Dr. Ravich, who is also associate professor of psychiatry at Cornell University Medical College, is how couples interact in getting both of these trains to the finish.

Each couple makes 20 runs per test, and after more than 15 years of train testing, Ravich has isolated nine patterns of behavior. In one, husband and wife alternate being the first to cross the finish line by using the shorter and faster of the two tracks. In another, the couple may simply decide to take different tracks, often an ominous sign for the marriage, according to Ravich. Couples who go their own ways in the game often go their own ways in marriage and may be prime candidates for divorce.

Ravich recently retired his toy trains in favor of video games in which each person manipulates a small blip of light. He began his long-term research project at the Marriage License Bureau in New York City, where he had newlyweds take his train test. Follow-up tests, done over a seven-year peri- od, show that, for better or for worse, couples' game playing patterns don't seem to change.

Pornography: If violent, it may create a cultural climate that encourages sexual assault.

BORN FIGHTERS There is some truth to the notion that people on the brink of death can survive if they possess an inherent "will to live." Indeed, injured mountain climbers or patients with chronic emphysema can avoid attacks by making a greater than normal effort to breathe. Not all people can muster the willpower. But those who do, according to a team of doctors at the University of Colorado, in Denver, are "born fighters," who have quite literally inherited the will to survive.

According to Dr. Robert Grover, who directed the research, the will to breathe comes from the tiny carotid sensor, buried deep within the neck. The sensor, he explains, consists of a group of cells located on the carotid artery (the vessel that supplies blood to the head). Whenever oxygen in the blood drops below a certain level, the sensor cells are activated and direct the respiratory muscles to work harder. However, Dr. Grover says, "the magnitude of the response varies tremendously from person to person. A sensitivie carotid sensor is something you're born with, something you get from your parents."

About 25 percent of Americans have low sensitivity, Dr. Grover says. But insensitivity is harmful only in circumstances that reduce the amount of oxygen in the blood. People who live in high altitudes, as well as those suffering from lung disease or obesity, could be adversely affected by a hard-to-activate sensor. Nevertheless, Dr. Grover has found that drugs that enhance breathing can endow the less sensitive with more tenacity than before.

VIOLENT PORNOGRAPHY Pigtailed and innocent, barely 18, she skips into the college dorm. Two young men sitting on a couch share a bottle of whiskey and grin as she flops down between them. Then the action starts: They shove her around and one forces liquor down her throat. When she protests, they throw her onto a table, tearing her clothes off and tying her arms behind her back. They beat her and rape her, and when she screams, they rape her again.

A weighty pane of glass . . . or just an illusion? Dartmouth students couldn't make the distinction.

The scene described is part of a film that plays over and over in the darkened laboratory of psychologist Edward Donnerstein, of the University of Wisconsin at Madison. Donnerstein is conducting a series of experiments to test the social impact of violent erotica. He has recently learned that when young men watch this depiction of rape, "they become incredibly aroused and angry." Their blood pressure soars 15 to 50 points, and when given the chance to punish female laboratory partners with noise, many become merciless, subjecting the women to the loudest beeps possible.

According to Donnerstein, the experiments suggest that violent pornography contributes to wife beating and rape. While it is impossible to trace an individual act of rape to a particular book or film, says Donnerstein, "the media's hostile depiction of women nevertheless creates a cultural climate that encourages sexual assault." The findings, Donnerstein adds, are the first to challenge the conclusions reached more than a decade ago by the National Commission on Obscenity and Pornography, which exonerated pornography as a cause of antisocial behavior.

GRASS INTELLIGENCE If you smoked 40 joints of potent Jamaican grass every day, visions of Day-Glo, or galaxies, would be dancing in your cerebral spaces, right? But for members of the Ethiopian Zion Coptic Church, near Miami Beach, Florida, pot, though a sacrament, is no more phantasmagorical than Tylenol.

Celebrants of this Rastafarian-inspired sect each day smoke two to four ounces (equivalent to about 40 joints) of the "green herb of the Bible"—tobacco-laced Jamaican marijuana. Astonishingly, though, their I.Q.s appear unaffected by seven or more years of steady, ceremonial consumption, according to UCLA neuropsychologist Jeffrey A. Schaeffer.

"We were rather surprised," says Schaeffer, a chemical toxicity expert. "Any episodic marijuana user is familiar with short-term memory defects, driving difficulty, and so on, but this group [he has refused to identify them,

102   BEHAVIOR AND THE MIND

> *"If the only tool you have is a hammer, you tend to see every problem as a nail."*
> *—Abraham Maslow*

although other investigators have] says they don't even get high. They don't giggle or get bloodshot eyes, either."

Not only have these seekers probably developed a tolerance to the drug, Schaeffer says, but their strict diet and routine austerities set them apart from the drive-in-burger and rock-and-roll crowd. Then, too, they use marijuana as an avenue to the Almighty.

The eight men and two women in Schaeffer's study, all young Caucasians, had an average of 13.5 years' schooling and a mean I.Q. of 128.

Robert Peterson, research director of the National Institute on Drug Abuse, in Bethesda, Maryland, cautions against drawing any conclusions from "such exceptional people. Not since Thomas Jefferson sat around talking to himself has any group had a mean I.Q. of one hundred twenty-eight."

Moreover, since the magic herb is mingled with tobacco, Peterson doubts that the disciples actually inhale much THC, marijuana's active ingredient. Schaeffer disagrees, noting that cannabinoids—metabolites of cannabis—were found in the subjects' urine.

The Coptic leader, Brother Love (Thomas Riley, Jr.), was tried for drug smuggling; he pleaded freedom of religion on the basis of his ministry's sacramental use of marijuana, but he was convicted anyway.

EASY BELIEVERS Recently a few Dartmouth College students ran through a snack bar, throwing fake snowballs at one another. Nearby two others carried a huge piece of plastic, puffing and panting as if they were burdened by weighty plate glass. When they all collided—on cue—yet another student produced sound effects by smashing a small piece of real plate glass to the floor. Then one of the bogus glass carriers broke a bag of blood-red ketchup under his arm, and the snack bar flew immediately into an uproar. A few minutes later casual witnesses were questioned: Though the students' observations varied considerably, they all declared that the incident appeared to be authentic.

Laundromat banter:
"Those are some nice undies
you have there"
is an unsuccessful opening line.

This drama, staged by Dartmouth psychologist William Smith, is only one part of a course on the psychology of credulity, developed to help freshmen improve their memory and avoid psychological manipulation. Toward this end, Smith's students stage and analyze events of this sort. They also try to understand the motives for believing con artists and pitchmen in an attempt to resist their message.

"What you believe is intrinsically bound to your psychological makeup and to what you already believe," Smith says. When people aren't ready for unexpected events, they miss a lot and let their memory fill in the blanks—with what they would have expected to occur. "My intent is to get the students to be more suspicious," he adds. "I don't want to turn them into cynics, but a good dose of skepticism never hurt anyone."

OPENING LINES What do you say to an alluring stranger? Well, if you're a man and the stranger is a woman, lines like "I bet the cherry jubilee isn't as sweet as you are" won't get you very far in Singlesland.

Men have underestimated just "how much women *hate* cute, flippant opening lines," says psychologist Chris L. Kleinke, of Edith Nourse Rogers Veterans Hospital, in Bedford, Massachusetts. In his study, both sexes—but particularly women—preferred either a "direct approach" or an "innocuous question" as a prelude to further acquaintance. Belittling sexual remarks proved highly ineffective.

Kleinke had college students rate a potpourri of opening lines appropriate to bars, restaurants, supermarkets, laundromats, beaches, and general situations. Highly rated lines included direct ones like "I feel a little embarrased about this, but I'd like to meet you" and innocuous queries like "What do you think of the band?"

The no-no's ranged from "Bet I can outdrink you" (bar) and "Is your bread fresh?" (supermarket) to such laundromat banter as "Those are some nice undies you have there."

Men employ such cutesy, macho come-ons out of fear of rejection, Kleinke theorizes. Indeed, a few women *liked* phrases like "Your place or mine?" (advocated in pickup manuals). But Kleinke recommends the ingenuous question as a more productive ego saver.

If you can't think of anything to say, at least say *something*. But don't talk too much, either. This advice comes from another Kleinke study, in which college students listened to tapes of a purported first encounter between a man and a woman and then rated the speakers on both their likability and dominance.

Whether male or female, the person who spoke 80 percent of the time was judged domineering, cold, and impolite, while the one who got in only 20 percent of the words came across as submissive. When the conversation was equally shared, both speakers scored high on likability, warmth, and politeness.

SNORING CURE "Laugh and the world laughs with you; snore and you sleep alone," goes an old saying. Snoring can be "a noxious interpersonal problem," says Raymond Rosen, a clinical psychologist at Rutgers University Medical School, in New Jersey, and he and his wife, Linda, an experimental psychologist, have patented a device designed to stop it.

The idea is fairly simple. When snoring exceeds a certain volume, the device wakes the patient up. It also counts snoring episodes during the night, to help make the patient aware that he really was snoring. During the two-week treatment period the device is set to respond to progressively lower snoring volumes.

So far over 80 percent of people using the device have reduced their snoring considerably, and many continue to improve even after the treatment is finished, says Rosen (who also is quick to note that neither he nor his wife snores). The device isn't on the market yet, but Rosen is talking with possible manufacturers who could market it for less than $100.

"The release of atomic energy constitutes a new force too revolutionary to consider in the framework of old ideas."
— Harry S. Truman

SPORTING VIOLENCE Every month or so John Cheffers lugs a pair of video cameras to his car and drives to Boston Garden to see the Bruins in action. While he's watching the game, he flicks on his cameras, pointing one at the hockey players and the other at the people in the stands.

Cheffers, a psychologist at Boston University, has for the past few years been studying the violence that erupts among spectators at sporting events, anything from hockey to football. Using data gleaned from video recordings and firsthand observations of more than 700 college and professional games, Cheffers has concluded that violence on the field or in the arena almost always provokes violence among the fans. At the Bruins' games, for instance, fans become rowdy whenever a brawl breaks out on the ice.

Because hockey players are especially violent, followers of that sport become more disorderly than any other sort of fan. Yet other sports also provoke rage. Soccer fans, for example, do not go to a game expecting players to slug it out, but when they do, the spectators often break into a slugfest themselves. Indeed, some 65 percent of all the fights among soccer players are matched by disturbances in the stands.

To keep peace among spectators, Cheffers would have TV scanners that can predict outbreaks. He would like to see fans grouped in smaller, "less anonymous" seating sections. If the chairs were more comfortable, the herd instinct that normally prevails in crowded stadiums could be subdued. And Cheffers suggests that arenas resemble parks, with flower boxes in the stands. Banning the sale of alcoholic beverages at athletic events, however, is one measure Cheffers does not recommend. "Most imbibers," he says, "are well behaved."

ODD COUPLES Are your bureau drawers full of neatly arranged, color-coded socks and shirts, or does your habitat more closely resemble Attila the Hun's rec room?

Freud traced such phenomena to toilet training and its attendant traumas,

Odd couple:
Now there are therapy workshops
to help couples
with great dissimilarities.

but New York therapist Selwyn Mills disagrees. It is your brain's inborn "perceptual style," he says, that accounts for your obsessive neatness, abominable slovenliness, or anything in between.

Researching what he calls the Odd Couple syndrome, Mills found that the finicky Felixes of the world are dominated by the brain's left hemisphere–the verbal, time-conscious, orderly, sequential half.

The brain's right hemisphere is imagistic, spontaneous, and holistic. The untidy Oscars, who squeeze the toothpaste from the top of the tube and leave the cap off, are right-brain people.

In addition, visual orientation corresponds to neatness, while "kinesthetic" people—who take in information predominantly through their senses, feelings, and intuition—are apt to be sloppier. Sloppiest of all are right-brain kinesthetics; neatest are left-brain visuals.

Potters, cabinetmakers, homebuilders, and other craftsmen, Mills notes, tend to be left-brainers. Artists take their cues from the right.

"The important thing to understand," Mills says, "is that neatness or sloppiness isn't intended to hurt anyone. It's just the way we're programmed. Sloppy people just don't *see* things; they don't perceive visual details."

Strangely, these mismatched opposites often attract one another, and MIlls has devised "Odd Couple" workshops to help them live together more harmoniously. "Eighty percent of the couples we've seen are composed of a tidy person and a messy mate. Neat parents also tend to have sloppy children, and vice versa."

PLASTIC BRAINS Four-year-olds can pick up Swahili or Urdu better than forty-year-olds because their brains are still "plastic," that is, crudely wired and susceptible to being changed by experience.

Alas, as we mature, our brains quickly become set in their ways. But California Institute of Technology researcher Takuji Kasamatsu may have uncovered the brain's fountain of youth.

Teen-agers' early sexual activity has been linked to early menstruation, a fact now being hotly debated.

Kasamatsu's experiments with visual perception in cats show that the neurotransmitter norepinephrine can return a mature brain to its youthful, flexible state.

In kittens the neural pathways for normal stereoscopic vision are fixed during the first few months of life. But Kasamatsu subjected adult cats to one-eyed vision by sewing one eyelid shut and then injected their brain's visual center with norepinephrine.

The result: The cats' brains were "imprinted" with the monocular experience as if they had been kittens. Norepinephrine had made their brains "plastic" again.

Could the neurotransmitter norepinephrine be to the human brain what health spas are to the body? "That's our dream," Kasamatsu confides, "to make your brain young again."

MENARCHE MYTH Teen-agers' sexual activity has increased dramatically in recent years. The accepted explanation: Today's girls simply reach sexual maturity sooner, menstruating at twelve and a half years of age, some four and a half years earlier than girls 100 years ago.

Now, however, it seems that this widely held explanation is incorrect. According to sociologist Vern Bullough, of the New York State University College at Buffalo, girls in nineteenth-century America began menstruating at the age of thirteen and a half, only one year later than present-day females.

The notion that nineteenth-century women did not begin to menstruate until the age of seventeen came from the books and articles of a British doctor named James Tanner, Bullough said. Not exactly a meticulous researcher, Dr. Tanner reached his conclusions after studying an isolated— and malnourished—group of Norwegian women. He did not realize that poor nutrition always delays the onset of menstruation, sometimes for years.

In the early 1800s other Norwegian communities recorded the onset of menstruation at fifteen years of age, Bullough adds; according to Roman,

Greek, and Arab sources, girls once began to menstruate between the ages of twelve and fourteen.

SLEEP/WAKE BIOFEEDBACK Skin conductivity, which reflects your level of alertness, has long been a key component of lie-detection analysis. Now Skin Conductance Response, or SCR, may prove useful for relieving insomnia or for preventing people from dozing off at undesirable times, thanks to research done at the University of Tokyo.

In one experiment, subjects were seated before a screen in a darkened room and were asked to press a button when shown a certain pattern of dots. This was very boring, and most subjects dozed off within ten minutes. The Tokyo researchers found that as the subjects fell asleep, their SCRs would drop below a certain level.

In the next study, the subjects who had dozed off were given the same task, but this time their SCR level was monitored. A bell alerted the subjects if their SCRs dropped too low. The technique worked well, and the subjects were able to perform their monotonous tasks.

In the final experiment, problem sleepers lay down in a darkened room and were instructed to decrease the volume of a recording of ocean surf sounds, using biofeedbacklike techniques. The volume was actually controlled by the subjects' SCRs, which they attempted to lower. Seven of nine insomniacs fell asleep.

Researchers Chiaki Nishimura and Jin-ichi Nagumo suggest that their findings "will be applicable to a doze-alarm system for drivers" and "to clinical therapies of sleep disorder."

EMOTIONAL ALARM CLOCK Hypnotic regression, the controversial therapy of reexperiencing earlier life traumas to obtain insights into psychological problems, has always had one complication. The therapist has not always known the best point to cease the regression.

Emotional alarm clock:
A New York psychologist is using
hypnosis and electrodes
to regress patients to exact times.

Now New York psychologist Dr. Ivan Wentworth-Rohr has come up with something that can do just that. He calls it an emotional alarm clock.

Dr. Wentworth-Rohr, chief of the behavioral therapy unit at St. Vincent's Hospital, in New York City, uses electrodes to monitor the most responsive body systems. The psychologist watches for changes in the activity of muscles or sweat glands, for example, as a patient is being regressed. When Dr. Wentworth-Rohr's machine senses a surge of activity in the monitored area, it sets off an alarm. Even if the patient is not consciously aware of it, he may have been regressed at that moment to a particularly upsetting time or episode in his past.

In one session, for example, Dr. Wentworth-Rohr had a patient imagine he was walking through his childhood home. The alarm went off when he reached the foot of the hall stairs. "This showed it was an emotionally loaded area," the psychologist declares. "Then I had the patient regress in time to an earlier age. The alarm went off at two years old. The patient remembered being left there untended and feeling isolated and lonely."

Dr. Wentworth-Rohr thinks his alarm might be vulnerable for other forms of psychotherapy as well. "There's a good possibility that your body remembers more straightforwardly than your mind."

SHOCK THERAPY In the movie *One Flew over the Cuckoo's Nest,* one of the punishments meted out to the patients was ECT—electroconvulsive therapy, or shock treatment. For 20 years psychiatrists have debated whether ECT helps cure depression. Radical psychiatrists, such as R. D. Laing, have said that ECT is used to control patients rather than to cure them. Research shows that people suffer memory loss following shock treatment. And now an experiment indicates that ECT is not even very effective.

A British team at Northwick Park Hospital, in London, divided 70 depressed patients into two groups. One group received real ECT, with the full shock delivered directly to the brain. The other group got simulated ECT;

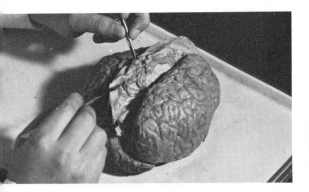

Even with half the brain surgically removed, patients were still able to perform a wide range of functions.

they were made to lie down on a special bed; electrodes were clamped to their heads; then they waited. But they were spared the shock. It has been argued that what helps patients may be the attention they are given before and after they receive the shock, not the jolt itself.

The patients were rated on three different scales used to measure depression. Both one month and six months after ECT ceased, the patients were rated again. During the course of the testing those patients who were shocked improved more than did those who were not. On the scale most used to determine a patient's degree of depression, the Hamilton depression scale, "real" patients improved by 38 points while "simulated" patients improved by 27 points.

The researchers point out, however, that in real-life terms such a difference is not all that great. Also, there was no difference between the scores of the two sets of patients either a month or six months after they received ECT.

Most psychiatrists in the United States and Great Britain view ECT as an effective means of treating depression. This new study gives opponents of shock treatment more ammunition for its curtailment.

HALF A BRAIN Half a brain is not only better than none, it may be just as good as a whole one, according to a report recently released by one British psychologist.

Writing in *New Scientist* magazine, Stan Gooch says that evidence coming out of operating rooms all over the world shows that the two halves of the human brain are not as specialized as some think and that the right side of the brain, in particular, has been underestimated.

According to what is known as the split-brain theory, the right half of the human brain, which controls the left side of the body, specializes in nonverbal mental work such as spatial or intuitive thinking. The left side, which rules the right half of the body and is considered dominant, does more logical work such as handling language or math problems.

How tall a child grows may depend on whether he lives in a relaxed or tumultuous home environment.

By this theory, losing the left side of the brain should cripple a person both mentally and physically. That, says Gooch, is not what happens.

His evidence comes from results of a drastic brain operation called a hemispherectomy, in which half of a brain is surgically removed. Patients in England and in South Africa who had the left half of the brain removed were still able to perform what are usually considered left-brain functions.

In one instance, a forty-seven-year-old man who had a hemispherectomy because of a large tumor on his left brain regained his use of speech and even sang old songs six months after the operation.

In another case, a twenty-one-year-old woman who had severe epilepsy and right-side muscle twitches because of a defective left hemisphere actually improved her so-called left-brain skills after a similar operation. Before the operation she did not read and had only sporadic control of her right side. After the operation she was reading books and newspapers for pleasure, had regained much of the use of her right side, and felt confident enough to take a job.

Gooch's explanation for what happened is that all intellectual and motor functions can be handled by either side of the brain. He suggests that as the brain matures, each side tends to specialize in certain functions but doesn't lose the ability to take over all if the need arises.

HOSTILE SHORTNESS Do happy and relaxed home environments produce taller children than homes in which the parents are hostile and cold?

A woman who had given birth to twins—a boy and a girl—and who, four months later, found herself pregnant again, was shaken up when her husband lost his job and left home. She began to take her hostile feelings for her spouse out on the young boy. Up to that time, the boy and the girl had been growing normally. But soon the boy's rate of growth plummeted behind his sister's. By his first birthday, he was only as tall as a seven-month-old. Later, when the child was placed in a hospital and his father had returned to his

*"The simplest schoolboy is now
familiar with truths for which
Archimedes would have
sacrificed his life."*
—Ernest Renan

mother, he began to grow. By the age of two, he had caught up to his sister's height.

It's called psychosocial dwarfism, a disorder that occurs when children are left in cold home environments and that seems to reverse itself as soon as they are placed in warmer surroundings. The reason it occurs at all seems to have something to do with the relationship between our hormones and the higher functions of the brain.

It appears that human growth hormone, which is secreted by the pituitary gland, located at the base of the brain, and which is responsible for most of childhood growth, is influenced by our emotions and our thoughts. Scientists have found definite connections between the pituitary and the brain. Chemicals called releasing factors travel down from the brain's hypothalamus into the tiny, pea-sized gland controlling when and how much growth hormone is secreted in the body.

So it is quite likely, say biologists, that these neural-hormonal connections play a role in how much we grow during childhood. And it is possible, they say, that children can, in fact, think small.

DAMP GENESIS Were our ancestors semi-aquatic apes, creatures virtually half-man, half-seal? Several recent popular anthropology books have argued that this is, indeed, the case. They state that our reactions to sea diving still mimic those of seals and we show the vestigial remains of a second eyelid we used for keeping our eyes clear under water. They also point out that the vast majority of human civilization lies along shores and riverbanks, a sign of deep human attachment to water.

According to Elaine Morgan in her book, *The Descent of Woman,* many human features are direct descendants of our watery period. She holds that about 12 million years ago some of our proto–ape ancestors returned to the water during extended droughts. In the shallows, these prehumans lost their body hair and began to stand erect and to use tools—to crack shellfish, for

Some anthropologists argue that we evolved from creatures that were virtually half-man, half-seal.

instance. The female hymen was developed to prevent the intrusion of seawater, and frontal sex became the norm, in the manner of dolphins.

Now, however, a report by medical physicist Jerold M. Lowenstein and anthropologist Adrienne L. Zihlman throws cold water on the aquatic-ape theory. They state that no evidence exists of a great drought during the Miocene era, the time we supposedly headed back into the surf. And, they argue, our ancestors lived deep within the central African plain and would hardly have migrated 2,000 miles for food without enormous provocation.

In addition, they note that human skeletons don't resemble those of aquatic mammals at all. The latter have much narrower pelvises for streamlined movement in the water and much shorter back limbs. Our skeleton, by contrast, seems designed specifically to bear heavy vertical loads without the added buoyancy of water.

And, the team adds, the human hymen isn't even waterproof. "What we are saying," Zihlman explained, "is that you can't make up theories out of the blue. You have to consider the constraints of science."

FORGETTABLE FACES The next time someone can't place your face, take it as a compliment. A group of California researchers has found that attractive people are often the most difficult to remember.

As part of a study on what makes people memorable, volunteers were asked to leaf through high-school snapshots of 120 white males. They then classified the photos in categories ranging from "usual-looking" to "very unusual." A second group of volunteers rated the same pictures on a scale from "least good-looking" to "very good-looking."

The men thought to be the most typical- (or usual-) looking were also judged the most attractive. In later tests, though, these were the same men volunteers tended to forget.

The reason, explains Leah Light, a cognitive psychologist who helped head up the study, is that we find people with typical looks attrractive. In the

Unforgettable face:
In a study of 120 white males,
the good-looking
ones were easiest to forget.

early 1900s, Sir Francis Galton first established a correlation between typical looks and attractiveness, Light points out. Light used Galton's work as a foundation for her own study.

But because these pretty faces generally lack distinguishing features—a large nose, crooked mouth—they're less likely to stick in our minds.

"If you want to be memorable," Light warns, "you'd better be a little ugly."

MARIJUANA-HEROIN LINK Many factors, including a victim's personality and environment, contribute to heroin addiction. However, scientists studying drug abuse are rekindling an old suspicion of a substance praised and promoted by its users for more than a decade: marijuana.

During the 1970s, the belief that marijuana whetted a smoker's appetite for the hard highs of cocaine and heroin was virtually dismissed. Marijuana use received discreet permission if not outright legal sanction: Even the New York State PTA called for decriminalization of the drug.

But now Wiliam Pollin, director of the National Institute on Drug Abuse, has stated that the stepping-stone theory "needs serious reevaluation."

After conducting a nationwide government-funded study of 2,510 men, sociologists John A. O'Donnell and Richard R. Clayton found that heavy users of marijuana frequently went on to try harder drugs.

The University of Kentucky researchers' survey included men who had smoked marijuana more than 1,000 times in three years. Almost three quarters of those men later tried cocaine, and one third shot heroin. On the other hand, of the men who'd used marijuana fewer than 100 times, only seven percent took cocaine. Four percent progressed to heroin.

While Clayton dismisses any biological link between marijuana and heroin, he's convinced that a "causal" link does exist.

THE MYTH OF MOTHER LOVE If you thought Joan Crawford unnaturally cruel to her daughter Christina, consider this: Nineteen thousand of the

117

21,000 infants born in Paris in 1780 were farmed out to wet nurses. Most of their mothers found breast feeding crude and child care tedious. Of those children who survived treatment by a nurse burdened with several charges, many did not see their real mothers until age four.

Likewise, in eighteenth-century France many parents chose not to attend the funerals of children under age five. They considered a younger child's death somewhat "insignificant."

From these examples and others throughout history, philosophy professor and author Elisabeth Badinter concludes that mother love may be more of a "gift than a given."

In her book *Mother Love: Myth and Reality* (Macmillan), Badinter questions such "eternal truths" as the universality of maternal love and that to feel fulfilled a woman must become a mother.

In Badinter's view, mother love fluctuates from age to age and from woman to woman. Time, place, and economic conditions all play a part in motherly affection: "The woman will be a good mother to a greater or lesser extent as a direct consequence of the level of esteem the society bestows upon the maternal role," explains Badinter.

The importance of being a good mother—and a veneration of the institution of motherhood—has slowly evolved over the past four centuries. As the child became more significant to society, so did the role of motherhood.

In the future, Badinter summarizes, mothering will be seen as a talent—rather than an instinct—shared by men and women alike.

OUR THIRD EYE Frogs have one. So do rats. But for years no one quite knew what to make of the tiny, pea-sized gland that resides also inside our own brain. Now it appears that the pineal gland may influence some of our most basic drives—like sex—and even some of our more advanced behavior as well.

In frogs the pineal is often called the third eye because it is sensitive to

In northern Finland, women are less successful at conceiving during the winter, when exposed to 24-hour darkness.

light. But the human pineal was thought to be just an evolutionary relic that, like our appendix, served no biological function.

Scientists, like Alfred Lewy, at the University of Oregon, have proved that belief wrong. They have discovered that when human subjects are exposed to bright light the pineal stops secreting its hormone, melatonin. Melatonin normally is secreted only during nighttime darkness. Because of this, Lewy and others believe that our pineals still function as light-sensitive "eyes."

Melatonin has been shown to have some powerful effects on the reproductive organs of animals. When bears hibernate during the winter, for example, their gonads shrink. This, researchers speculate, may be due to the suppressing effects of the pineal hormone on sex glands during prolonged darkness.

The same may be true for humans. Women living in northern Finland, the land of the midnight sun, are more successful at conceiving during the summer, when they are exposed to 24-hour sunlight, than they are during the winter months, when round-the-clock darkness prevails. Also, infertility is more common in blind women, which seems to indicate a connection between the pineal, the eyes, and the sex organs.

Just as interesting is the pineal's relation to the phenomenon known as jet lag. Lewy and others believe that the pineal is in some way responsible for resetting our clocks when we travel across time zones. But it needs stimulation from sunlight or bright artificial light to do the job well. Those who suffer from jet lag may not be getting enough sunlight after a long flight through several time zones.

LIFE IN THE WOMB Are babies in their mothers' womb awake? For generations people have known that fetuses kick and move. More recently doctors have determined that they breathe, inhaling and exhaling the amniotic fluid. Ten years ago, Geoffrey Dawes performed an experiment that indicated womb dwellers might just be awake. Fetal lambs removed from the

> *"I have hardly ever known a*
> *mathematician who was capable of*
> *reasoning."*
> — *Plato*

womb, Dawes said, sometimes opened their eyes and moved their heads. Occasionally they even showed signs of animation.

But Dawes's test was far from conclusive. Recently scientists at the University of Manitoba and the Nuttfield Institute for Medical Research in Oxford performed a more decisive test.

They implanted electrodes in fetal lambs, leaving them in the womb. They tested the fetuses to see if certain reflexes that are stimulated during adult lamb wakefulness would also increase during the fetuses' active period. They found "a very strong indication" of wakefulness in the fetal lambs and feel the same is likely in human babies.

Why would a fetus be awake? No one is certain, but one possibility is that it needs to become conscious of breathing so that it can cope with the process after being born.

DOMINICAN SWITCH Are the brains of men and women different?

Yes, according to endocrinologist Julianne Imperato-McGuinley, of Cornell Medical College in New York City. Imperato-McGuinley's evidence: 38 men in an isolated part of the Dominican Republic who started life as girls. At the age of eleven, when normal girls begin to show breast development, the 38 showed no change. At twelve, most of them began to feel the stirrings of sexual desire for girls. At puberty, their voices deepened, their testacles descended. and their clitorises enlarged to become penises.

These children came from a group of families carrying a rare mutant gene. The gene deprived them of an enzyme needed to make the male hormone testosterone work in the skin of their genitals. For this reason, their external genitals looked female at birth. But at puberty their bodies were able to use testosterone without the enzyme. It became obvious that they were males, as chromosome tests confirmed.

All but two are now living with women. They have male musculature, and although they cannot sire children, they can have sexual intercourse. They

Normal Dominican girl:
Thirty-eight unusual subjects
from her country
made a dramatic switch later in life.

have assumed masculine roles in their society.

This group was able to adjust, concludes Imperato-McGuinley, because in the girl's body was a male brain, virilized by testosterone before birth and activated by another rush of testosterone during adolescence. "To the world," she says, "they looked like girls when they were younger. But emotionally, they had always been male."

POTENTIAL RAPISTS Can any man become a rapist?

To find the answer, psychologist Neil Malamuth, of the University of Manitoba in Canada, asked hundreds of Canadian and American college students whether they would be likely to rape if they knew they wouldn't be caught. More than 50 percent said they might, and 35 percent said they probably would. Malamuth found that this 35 percent was far more likely than other men to believe that women enjoyed rape. They were also likelier to be sexually aroused by rape scenes in movies and books.

"Many people in the general population have attitudes similar to those of the convicted rapist," Malamuth concludes. It seems probable that some rape occurs when these people are pushed over the edge by forces in society. These "forces," Malamuth explains, include the recent surge of sexual violence in pornographic novels and films. Also at fault are highly regarded mass-market films like *Swept Away,* which hasn't a single nude scene but which shows a woman enjoying systematic rape and torture.

RADICAL AUTISM THERAPY The child rocks back and forth and stares vacantly into space, never responding to his parents, who are calling to him. He is not ignoring them; he simply does not notice that they exist.

The child is a victim of autism, a poorly understood disease whose victims have little awareness of external reality. Most autistics are self-destructive, make little eye contact with anyone, and communicate only through a few random sounds. After years of studying the disorder, therapists have not

121

The athletic brain:
Basketball can change the brain's
structure and promote
skills in spatial reasoning.

been able to break through. But a new therapy based on simple physical stimulation promises to bring some autistic children into the world of reality.

Developed by Brooklyn, New York, psychologist Ezra Gampel, the therapy consists of a series of daily half-hour sessions in which the therapist grips the autistic child and vigorously rubs his head, face, and body until the youngster begins to scream. After several months of this treatment, says Gampel, the autistic child begins to interact with parents and friends.

Normal infants, Gampel explains, usually begin to respond to the world around them at three months of age. Autistic children, on the other hand, never learn to react. The reason may be that stimulation seems hundreds of times more powerful to the autistic child; because even ordinary contact causes him to experience a sensory overload, he withdraws from the world to protect himself. With Gampel's therapy, however, the child is forced to deal with extreme and systematic physical stimulation.

So far, Gampel has successfully treated eight autistic children; some of them can now make eye contact with others, feed themselves, and even talk.

ATHLETIC BRAIN Can participation in athletics change the structure of the brain? Yes, according to Anne Petersen, director of the Adolescent Laboratory at the Michael Reese Hospital and Medical Center in Chicago.

In a recent study, Petersen found that boys who excel in athletics also excel in spatial reasoning—a skill controlled by the right hemisphere of the cerebral cortex and defined as the ability to understand maps and mazes or objects rotating in space. Says Petersen: "An athlete must be constantly aware of his own body and a whole constellation of other bodies in space." A daily game of basketball, she explains, might stimulate the secretion of hormones that prime a player's brain for success in basketball.

"Women are far less athletic than men," adds Petersen, "and also far less adept at spatial reasoning Perhaps some women just never develop the area of the brain specialized for spatial control."

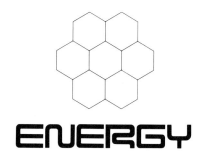

ENERGY

CHAPTER 5

A train on a circular track, underground, would store energy during the night to be used during the day.

ENERGY TRAIN Imagine a train burrowing deep under Paris or Rome at 400 mph. Floating on a bed of magnetism, this speedy vehicle will carry a quarter of a million tons of concrete instead of passengers, and it will never stop. It is intended to provide 100,000 families with electricity during the busiest hours of each day.

The train is called KRESS, for Kinetic Ring Energy Storage System, and if some British physicists are correct, it will soon be generating power for people around the world.

Mike Russell, of Rutherford Laboratory, in Didcot, England, explains: During the night, when utilities produce more electricity than anyone can use, they will funnel it into the motors that propel the train in a circle. As the train glides with ease along the layer of magnetism, its huge concrete mass will literally "absorb" the motion and store it in the form of kinetic energy. In the daytime, when consumers require *extra* electricity, the utility will flip a switch that converts the motors into electrical generators. The train will slow down as its store of kinetic energy turns the wheel of the generator, providing an entire city with power.

Today most utilities store power with the help of hydroelectric systems: At night they use their extra energy to pump water up a mountain, where it is stored behind a dam. During the day they open the dam, and as the water flows back down it turns a generator that produces electricity. The chief problem with this method is that you need a mountain handy.

Russell believes that KRESS will be far more practical for almost everyone. The British researchers are studying the economics of the system; they hope to build a working model in about three years.

FUEL-SAVING WINGS Engineers for the National Aeronautics and Space Administration (NASA) have tested a new wing attachment—called a winglet—that should cut airlines' fuel bills.

Viewed from the front, an airplane equipped with winglets looks like the

The Flying Nun used
winglets on her hat. NASA believes
a similar device
will cut airlines' fuel bills.

head of a longhorn steer. The metal winglets tested on a commercial DC-10 stand ten and one-half feet tall at the wingtip. A smaller device attaches underneath each tip.

More important than its novel appearance are the fuel savings. NASA engineers claim the energy savers can reduce fuel use by 3 percent on the DC-10s. That amounts to an annual savings of 250,000 gallons of jet juice, at a current cost of just over a dollar per gallon.

How do the winglets work? By lessening the friction (drag) caused by air passing along the airplane's wingtips. These whirlpools of air that swirl off the tip of each wing account for 40 percent of the total drag on an airplane flying level. Winglets cut down the energy needed to keep the plane airborne.

Currently, winglets are being fastened to private jets, such as the Gates Learjet 55. Commercial Airlines have yet to try the design on their own.

In fact, NASA's invention could have a hard time getting off the ground. The world's largest producer of commercial airliners—Boeing—froze the wing design of their newest models several years ago, before the effects of winglets were discovered. So no one is certain of how many airliners can be successfully adapted to winglets.

ITALIAN TOWER OF POWER Adrano, a tiny hamlet in Sicily on the slopes of Mount Etna, is playing host to an electricity generator that will upset no environmentalists. For at Adrano the European Economic Community and Italy's electricity utility, ENEL, have started testing a solar power station that can produce 1 megawatt of electricity.

The plant, called Eurelios, is of a type known as a power tower. Fields of focused mirrors are angled to reflect the sun's rays into a specially designed boiler. As the earth rotates, the 182 mirrors are moved by an electronic control system to keep the reflected energy pointing into the boiler. The boiler produces steam to drive a turbine linked to a generator.

Eurelios is the first of five similar designs around the world to go into action, and it is the first solar power station in Europe to be connected to a national electricity grid. Two of the four others are in the United States, one is in France, and the fifth is in Spain.

Although the energy is free, the cost of building and running the plant is high. For Eurelios, an experiment that will probably never be repeated or be taken up by industry, the cost is 20 times that of an oil-fired power station. Instead, scientists and energy administrators are looking toward solar photovoltaic cells for the future to produce electricity directly from the sun. The cost of these is also high—over ten times the price of an oil-fired station—but recent developments in the United States with thin films of silicon are raising hopes that it will come down fast.

PLANT PETROL Are the OPEC oil-price hikes getting your down? Do gas station attendants snicker when you drive up to the pump? If so, cheer up. This may not be for long. Nobel Prize–winner Melvin Calvin and his associates at the University of California at Berkeley have discovered a group of plants related to the rubber tree that can make their own petroleum.

Four years ago, Calvin found in Brazil a plant, of the genus Euphorbia, that produces the same kind of sap that rubber trees do. Rubber, which is actually a latex made up of oil and water, is jellified in the commercial process. "Only in this case of this plant," adds Calvin, "when you get rid of the water, you're left with oil."

Euphorbia is an ideal energy supplier because it is found not only in Brazil but in just about every country on Earth. Calvin decided to use the plant species native to the American Southwest simply because "it happens to suit the area." But, he says, any of the other species in the group would also be good fuel makers.

At this early stage, the researchers are observing how the plants hold up to their new job. So far, a stand of Euphorbia growing on Calvin's California

Oceans cover 70 percent of the earth. They are now being harnessed for energy by exploiting temperature differences.

homestead is producing about ten barrels of petroleum per acre at a cost competitive with world prices. And, with selective breeding for increased sap production, it is possible that the plants will yield even more hydrocarbons. "After all," adds Calvin, "twenty years ago we were growing corn, and if we got forty bushels per acre we were doing pretty well. Today the average is a hundred bushels per acre."

Does he plan to reap the bounty of his discovery? "I sure do," he says. "As soon as somebody makes enough of it, I'm going to put it in my car."

OCEAN ENERGY The oceans, which cover 70 percent of the earth's surface, have absorbed and stored sunlight for eons. Unfortunately, the solar energy in this vast reservoir has never been accessible to man. Now, for the first time, scientists are learning to recover large quantities of this energy with a technology called OTEC, for Ocean Thermal Energy Conversion.

The concept of OTEC was spelled out a century ago, when French physicist Arsène d'Arsonval suggested that energy could be generated by exploiting the temperature difference between the ocean's sun-warmed surface and its chilly depths. The viability of D'Arsonval's idea was demonstrated last year, when the Lockheed Corporation launched a vessel called *Mini-OTEC*. *Mini-OTEC* produced electricity by drawing warm water from the surface of the ocean into a chamber containing liquid ammonia (which has a very low boiling point). The water temperature was high enough to boil the ammonia, turning it into a vapor that rushed past the blades of a turbine. The turbine began to spin, producing electricity, and the ammonia vapor then flowed into another chamber cooled by 40°F water pumped up from the ocean's depths. At this temperature, the ammonia gas condensed back into a liquid and was ready for another round in the cycle.

By boiling and condensing the same ammonia again and again, *Mini-OTEC* produced a mere 50 kilowatts of electricity, just enough to run its own equipment, including a few floodlights. But it was enough, according to Rob-

Computers, CAT scanners, and satellites are slowly replacing exploratory oil-drilling rigs.

ert Cohen, of the former Department of Energy, to convince a few skeptics that OTEC could work. The DOE completed the conversion of a tanker into a small test plant dubbed *OTEC-1*. Now anchored 18 miles off the coast of Hawaii, the department will soon begin testing OTEC technology in preparation for large pilot plants to follow.

PROSPECTING WITHOUT DRILLS One of the biggest expenses behind the high price of oil is the enormously costly process of prospecting. Punching holes in the ground in the search for oil gobbles up oil-company profits, ensuring that they will have enormous outlays to figure into the price of any eventual find.

Now geologists are turning to technology to help them find new oil sources without having to dig. Computers and satellites are doing what tromping around hills with a rock hammer once did.

Using systems much like the medical CAT scanner, oil prospectors are taking detailed three-dimensional pictures of the earth's interior. These real-world maps have vastly reduced the guesswork of relating surface features to the underground presence of black gold.

Computers—including a system called Prospector, which emulates the questioning technique of a human geologist in a fashion that borders on artificial intelligence—have changed the face of geology. Storing seismological information is the second-largest use for computers today. In the near future, still larger and faster computers will store data from microprocessor-controlled radio sensors in the field.

Satellites are sending back increasing amounts of information to help geologists create a clearer picture of the world's subsurface. Currently used satellites have provided detailed images of the earth's magnetic field and gravitational pattern, important clues to what lies down deep. A new satellite will soon begin analyzing Earth's heat radiation, offering a whole new area for study. By comparing heat features of known deposit areas with those of potential sites, geologists hope to pinpoint the best prospects without leaving their offices.

No high-tech tools can eliminate the need eventually to drill a hole and see if oil is where you think it is, but, says an industry executive, "this new technology will give us a more accurate rendition of what's down there."

SOLAR MOBILE HOME When Ted Bakewell III pulls his mobile home into a trailer park, he doesn't hook it up to the local electricity supply. He doesn't tie in to the park's sewer system. His mobile home, which looks like a hyperthyroid lady bug, is completely self-sufficient.

Bakewell, thirty-five, owns a company that manufactures solar panels and builds passive solar buildings. He needed a portable dwelling he could plunk down at construction sites that were often quite remote from conventional power sources. So he and designer friend Michael Jantzen put together the Autonomous Dwelling Vehicle, the world's first self-sufficient solar mobile home.

Electricity in the new-age trailer comes from photovoltaic panels that cover the front wall and a small wind generator that stores up electricity while the home is being towed. Water is collected from the rain and is heated by passive solar collection and the waste heat from the trash incinerator. Human waste and other organic material are composted in a Clivius Multrum, a waterless, chemical-free, self-contained Swedish toilet.

"There is no energy crisis," Bakewell states. "There is a design crisis. The universe is nothing but energy. We have to find ways to harness it. The big question is, How do you reduce dependence on fossil fuels without living like a caveman? This house demonstrates there are alternatives."

ENERGY MOUNTAIN In Britain the Central Electricity Generating Board (CEGB) has found a way to level the peaks and valleys of power demand by using a peak of its own in the Welsh mountains.

129

The idea is simple as
a mountain waterfall. Pump water
uphill by night, let
it flood back down by day.

At Dinorwic, a craggy promontory in the Welsh coal country, the CEGB has excavated 1 million cubic meters of earth to create a cyclopean network of tunnels and rooms running through the hills. This subterranean causeway will generate 1,320 million watts in 10 seconds, using no fuel and creating no pollution.

The principle behind the system is as simple as a mountain stream—water running downhill. At Dinorwic CEGB plans to pump water to a reservoir at the mountaintops during nighttime hours, when electrical demand is low and conventional power plants have excess working capacity. Later, when demand soars beyond the capabilities of the workaday power plants, CEGB can let loose the mountaintop flood through Dinorwic's waterways and create an instant burst of power that can last for up to five hours without diminishing.

The CEGB estimates that Dinorwic will save $80 million a year when it goes into operation late in 1982. And it should keep on saving energy almost as long as the mountains remain standing—and gravity stays in effect.

SOLAR AIRPLANES The delicate airplane skidded down the runway at Edwards Air Force Base and rose into the California air. The translucent craft, named the *Gossamer Penguin*, was powered solely by the sun.

Piloted by 95-pound Janice Brown, the plane flew for two miles at a speed of 15 mph and a top altitude of 12 feet. All Brown had to do besides steer was keep a set of solar panels facing the sun. The panels, 50 square feet of "photovoltaic cells" capable of converting sunlight into electricity, powered a motor, and the motor, in turn, drove a propeller that pushed the craft through the air.

The recent debut of the 68-pound *Penguin*, says its builder Paul Mac-Cready, was just a prelude to future solar-powered flights. Indeed, Mac-Cready is already planning the first flight of the 110-pound *Solar Challenger*, a plane equipped with 30,000 solar cells and capable of flying 100 miles.

*"The development of
hydro power in the desert of
North Africa awaits only the
introduction of water."*
    —Statement in Nuclear News

MacCready does not envision commercial solar flight. "I want only to capture the public's imagination with the viability of a particular energy source," he says. He has already captured the imagination of NASA officials, who speculate that solar planes may someday be used for anything from studying the earth's resources to spying.

HOT ROCKS The foundation behind Minneapolis-St. Paul's new energy-saving scheme is the actual foundation beneath Minneapolis and St. Paul. University of Minnesota geologists are thinking about turning certain areas in the bedrock upon which the cities rest into huge subterranean energy-storage vaults. If the plan were put into effect, the Twin Cities would sit atop their own vast energy bunker.

Beneath the cities, geologists have determined, lies a layer of porous rock, in which water and impermeable rock are mixed. Such areas are called aquifers because of their water-holding ability. Four of these geologic sponges lie beneath the Twin Cities.

The Minnesota Geological Survey is especially interested in the Franconia-Ironton-Galesville aquifer, which rests some 700 feet below the city surface. Geologists conceive a system in which two wells would be drilled into the aquifer. One would pump cold water out. The water would be passed through a heat exchanger and this heated water would be pumped back into the aquifer by the second well.

Over time, the aquifer water would get hot, and, because of the enormous thermal mass of earth that surrounds it, it would retain 95 percent of the heat for up to six months. The stored heat could be tapped whenever needed.

"The beauty of this operation is that it is a nonconsumptive use of water in the aquifer," says Matt Walton, of the Geological Survey. Also, it would allow off-season heating capacity and coal at off-season prices to provide heat for buildings during Minnesota's long, cold winters. "This is one of the quickest methods you can use to save a large amount of energy," Walton states.

Alternate energy or tax loophole? Actually, both—water wheels and other devices now qualify for tax breaks.

LEAVES OF GAS If the resource-parched future is to sustain the world's technological society, clever technologists are going to have to figure out new ways to use wastes from existing processes. A recent test at Texas A&M University shows the potential of such developments—in spades.

If there are two things the southern United States has too much of, it's sawdust and bark. The South's hills are thickly covered with lumber pine, but even the most efficient milling operation produces some sawdust. And there are few important uses for the tree bark.

At Texas A&M, however, researchers used these waste products as the basis for a process that creates a hydrocarbon compound almost identical to the essential ingredient in gasoline and diesel fuel. By using the tree products instead of crude oil and "refining" them in a modified version of petroleum distillation, the scientists created a product that looks, and acts, like engine fuel.

When they tested their leafy power fluid in a stationary engine, it worked as well as the gas it replaced.

Now the Texas A&M team plans to try similar experiments with corncobs, rice hulls, and cotton-gin trash. One can even imagine a day in the future when our garbage goes in one side of a home converter and gasoline comes out of the other.

ENERGY TAX SHELTERS Tax shelters come and go, but money in search of a loophole always finds a way to get off the tax files. The latest fad among the fast money finance set is tax shelters for alternative energy development. If you have the scratch, you can save yourself tax money by investing in anything from wood chips to home water wheels.

The current avaricious interest in new power sources stems from Congress's attempts to come up with answers to our foreign-oil problem. It passed bills creating hefty windfalls for those willing to risk their money on experimental energy schemes.

As a result, a number of peculiar investment outfits have arisen. An example is Windfarms, Ltd., a California company that is beating the bushes for investors to fund two $4.5 million wind-power plants in Hawaii. They entice investors with the lure of an eight-year payback on their money.

Water power is another big area for the new tax shelters. One report said that "promoters all over the country are poking around rural areas, looking at fast-moving streams and long-abandoned dam sites," hoping to find potential hydroelectric locations.

Even good old wood is getting a shot in the arm from the windfall investment. Greater Boston Development, Inc., a small investment banking firm, is setting up syndicates to buy plants that produce something called densified wood for use as boiler fuel.

The good thing about all this is that it may stimulate some new energy ideas. A negative possibility is that all the fast-buck action may well create a spate of bad projects that will lose people's money and tarnish the alternative-energy movement for years to come.

HOT JELLO What's flat, wiggly, 16-feet across, and can heat your home's water? A gel pond, that's what.

In the hills outside Albuquerque, New Mexico, scientists from the University of New Mexico have designed and built a variation of the solar pond that may actually make environmental solar engineering practical in some situations. Basically, what the team—led by chemical and nuclear engineering professor Ebtisam Wilkins—has done is wrap a solar pond in gelatin.

Solar ponds contain a layer of saltwater with a layer of freshwater on top. Sunlight passes through the upper layer and is trapped—the greenhouse effect. However, since the saltwater is heavy it can't rise to the surface when it warms, as ordinary water would. It gets hotter and hotter in the depths, until it becomes warm enough to be pumped past a heat exchanger and heats up home water.

Garbage dumps no
longer just fuel hungry birds.
Landfill methane is
now being "mined" for energy.

To create a gel pond, a layer of transparent polymer gel is placed across the water's surface. This eliminates virtually all heat loss, making the pond much more efficient and practical as a heat source. It also cuts down on evaporation, a serious headache with uncovered ponds.

Wilkins states that a pond 16 feet in diameter and just 4 feet deep could provide all the hot water for a 2,000-square-foot home. He also believes the hot water from the pond could turn a turbine and generate electricity.

ENERGY FROM DUMPS Billions of microbes in a New Jersey landfill are tirelessly converting garbage to methane gas. The gas fuels just a few furnaces and machines at the Hoeganaes Corporation, a firm that manufactures iron and steel powder. But within the decade, says New Jersey's Public Service Electric & Gas Company, methane produced in this landfill and others could supply the northeastern United States with large quantities of energy. Methane is the main constituent of natural gas.

According to engineer Douglas Nielsen, of Public Service, most landfills are chock-full of micro-organisms called methanobacteria, which digest refuse and emit methane as a waste product. If the methane is left alone, says Nielsen, it eventually seeps out and may explode or erode local soil. But if it is collected in wells, this same gas can fuel homes and factories.

Public Service is currently trying to learn how much fuel each kind of garbage can produce. It is also studying how methane production varies with rain, snow, or extreme cold. "We still don't know about the basic nature of a landfill," Nielsen says. "But once we learn which factors speed up the production of methane, and which slow it down, we'll be able to offer our customers a dependable source of fuel."

GARBAGE-POWERED CARS Garbage-generated methane can also be used to run motor vehicles.

Waste gas from an Oregon landfill that once generated complaints from

135

Gas from a landfill is running this generator, which in turn is powering the lights on the Christmas tree at top.

citizens may soon power cars and trucks instead. Roughly half of the thousand cubic feet of gas the landfill produces each minute is methane, and purification of this gas for use in vehicles has been proposed by Emcon Associates, Inc., a San Jose, California, consulting firm that studied the problem for Clarkamas County, Oregon, officials.

Natural decay processes in the airless environment of a sanitary landfill produce a mixture of gases that is about half carbon dioxide and half methane. Methane, although odorless, can be hazardous. More offensive to landfill neighbors, however, are small quantities of such noxious gases as hydrogen sulfide, which smells like rotten eggs.

In 1978 a gas-collection system was installed at the Oregon landfill to control unpleasant odors, but county officials soon realized that the methane they were burning could be used as an energy source. One possibility was feeding the methane directly into a natural gas pipeline. That is done in a Mountain View, California, landfill, but, according to Marya Donch, an engineer at Emcon Associates, the level of purification required is too costly for the smaller Oregon landfill. Instead, the consulting firm recommended using the methane to run cars and trucks, which don't require purification that's as extensive.

Some 20,000 vehicles in the United States have already been converted to run on natural gas rather than gasoline, at a cost of $1,300 to $1,500 each. Clarkamas County plans to add a few hundred to that total. However, the main reason for shifting to natural gas has generally been to reduce vehicle-exhaust pollution—not to burn up pollution from other sources.

TECHNOLOGY

CHAPTER 6

In the robotic future anyone who wants to pitch for his local team will simply borrow the memory of a pitcher.

ROBOT EVOLUTION Taking up where Charles Darwin left off, a scientist at Carnegie-Mellon's Robotics Institute has linked the future of human evolution to robots and the computer. Within 30 years, according to Hans Moravec, computers will be more intelligent than humans, and robot limbs will be more durable than human ones. The result: People will trade their feeble mortal frames for powerful robot shells and their plodding brains for superintelligent computers.

At first, Moravec says, people will try to preserve individuality, transferring exact replicas of their brain patterns to computer programs. Processed by speedy computers, the programs will enable people to think thousands of times faster than before and will provide them with remarkable flexibility. Robot people could, for example, transfer brain programs from shell to shell. Thus, if you wanted to conduct research on a distant planet, you could simply have someone build a robot on that planet's surface, then radio a copy of your program over.

The next evolutionary leap, Moravec continues, will occur when superhuman robots give up their desire for individuality and begin to share programs.

If, for example, a writer wanted to build a cabinet, he or she could save time and energy by borrowing the memory of a carpenter. The carpenter, in turn, could borrow the memory of a baseball player before pitching for the neighborhood team. And hundreds of scientists with access to minds as brilliant as Einstein's would spend their days fathoming the universe—and improving the race.

As people select and discard memories freely, Moravec says, the concept of self will be blurred. Moravec envisions the ultimate merger of all human brains with the brains of other forms of life, both earthly and extraterrestrial. After years of memory exchange, he says, we might well wind up with a single conscious entity whose memory is stored in a vast bank spanning the universe.

THE AIRMOBILE With oil prices skyrocketing and alternative fuels failing to live up to their ballyhooed promise, it sometimes seems as if the only fuel we'll soon be able to use in our cars is air.

No problem. A new automobile engine designed by R & D Associates, a development company in Marina del Rey, California, will actually run on air. The company notes that an air-fueled car would operate free of imported fuels, would produce few pollutants, and would cost the same to operate as today's vehicles. It even claims air will provide superb acceleration.

How does it work? Electricity from an ordinary home current is used to chill room air until it liquefies. The liquid air is stored in an insulated tank that is carried from the house and stashed in the engine.

There, a tiny boiler, heated by small amounts of propane, changes the liquid air into a hot, high-pressure gas. The gas drives the engine much as steam powers a steam engine.

The company doesn't say so, but its engine raises the interesting possibility of being able to reinflate your tires with the same stuff that runs the engine. A prototype should be on the road by early 1983.

SYNMETALS Electrical wiring for airplanes made of graphite. Custom-made superconductors that work far above the temperature of liquid helium. These are just two of the potential applications of synmetals, new materials being created by scientists to blend the properties of metal and organic compounds.

As metal becomes scarcer and the demands of high-tech uses more rigorous, natural materials fall short as future sources of electrical wire and other nonstructural products. So scientists are working to create substances that would take their place or improve on their performance. Many in the field believe the development of synmetals will be the biggest thing in science over the next 20 years, as important in its way as the spread of silicon semiconductors was 20 years ago.

As new airplanes strive to go faster, and the price of copper rises, graphite wiring will become more important.

An example of synmetal work is experimentation with graphite electrical wire. Normal graphite doesn't conduct electricity, but it is far lighter, cheaper, and abundant than copper. Researchers have found that by loading the graphite with such chemicals as arsenic pentafluoride, they can make it conduct electricity. The work is still highly experimental, but looking ahead, a Department of Defense scientist says, "When you think of the miles of cable in the wings of an aircraft, the change from metals to graphite means a tremendous savings."

As for superconductors, French scientists recently showed that a kind of synmetal called a charge-transfer salt can become superconducting at very low temperatures. Because the compound is synthetic, there is hope that tinkering with it could make its superconductivity much more potent. Creation of a custom superconductor is years off, but researchers believe charge-transfer salts may be the basis for a nonresisting electrical component that works at 77° Kelvin, four times higher than existing superconductors. It could be cooled by liquid nitrogen, rather than helium, a major step forward, as liquid nitrogen is the cheaper and more plentiful of the two.

As fledgling a science as it is, synmetal proponents are convinced they are on the wave of the future! "It may take ten years to develop the first real commercial application and twenty years before it becomes part of daily life," says Gerhard Wagner, a physics professor at Germany's Freiburg University, "but I'm absolutely convinced we're moving in that direction."

ELECTRIC PLASTIC The material used to make modern fish nets may in a few years bring us appliances with no motors and table tops that function as speakers or digital display boards.

PVF_2—polyvinylidene fluoride, as the material is called—became common in fish nets because it was a cheap, strong plastic that blended well with sea water. Then a Japanese professor discovered that when the plastic was heated and stretched it was transformed. PVF_2 became piezoelectric—it

> *"The energy produced by the breaking down of the atom is a very poor kind of thing. Anyone who expects a source of power from the transformation of these atoms is talking moonshine."*
> —*Ernest Rutherford, 1933*

could conduct electricity and move when exposed to a current.

Heating and stretching, research shows, lines up the polymer molecules of the plastic into pleats that polarize the material. When the plastic is placed between thin metal films and exposed to electricity, some of these pleats expand and others contract, causing the plastic to move.

This property caused PVF_2 to become popular in speakers, microphones, and other audio components in which electricity and motion combine. Pioneer uses the plastic in its products.

But far more important is work being done by RCA in Japan. There, PVF_2 strips are arranged like a thermostat, with one side expanding and the other contracting when electricity is present. This bending "converts electrical energy into mechanical motion with 30 to 50 percent efficiency," according to Ed Johnson of RCA's Tokyo lab. This is far better than a traditional electrical motor and raises the possibility of building appliances in which the materials and the motor are one and the same thing.

Already, RCA has built a motorless fan; electricity causes PVF_2 fan blades to move swiftly back and forth, stirring up a stiff breeze. Many other possibilities loom. Piezoelectric plastic motors would have no moving parts and would need no lubricants. They would suffer none of the mechanical energy waste of current motors. And they could "work in an environment that would kill a motor," according to Johnson.

The electrical plastic can also be used for digital displays such as those in calculators. A PVF_2 table top, shielded by glass, could print out instructions for cooking or the morning news. Piezoelectric blinds could raise, lower, or close themselves, depending on how much sun there was.

The potential uses are innumerable.

BLINDSIGHT "Pole, eleven o'clock, eight feet."

That could be the mechanically spoken message to a blind person wearing a device that warns about obstacles and gently taps the wearer to indi-

*"Given plenty of time, there
are few limits to what a
technological society can do."*
            —Freeman Dyson

cate the direction and distance of the obstacle.

A prototype of the system is being tested by Dr. Carter G. Collins, a biophysicist, and Michael F. Deering, of the Smith-Kettlewell Institute of Visual Sciences, in San Francisco, with research funds from the National Science Foundation.

A TV camera on the blind person's shoulder produces an image of the path ahead. This information is instantly translated by microprocessors into machine-generated speech about obstacles or landmarks and delivers gentle taps on a belt the person wears. The taps clue the user to the direction of the obstacle, one tap for something distant, more taps as the object comes closer and closer.

"The cutting edge of electronic technology is being utilized," says Dr. Collins. The prototype is bulky, but an easily portable system, performing more recognition tasks, is the goal, Dr. Collins says. "We've already constructed, debugged, and demonstrated most of the hardware and software for the present system, but the more sophisticated, miniaturized system will need further development."

SUPER SPERM If there is a faster sperm, Alan Barr will build it, or at least design it.

For the past few years mathematician Barr has been mixing biology, computer graphics, and analytical math as part of his research for his doctoral dissertation, "Spermatozoan Head Shape: A Theoretical Analysis," which he is now completing at Rensselaer Polytechnic Institute, in Troy, New York.

He has been experimenting with different head and tail designs that he distills into mathematical equations and feeds into a computer. The computer takes the numbers and "draws" them as animated sperms on a TV screen. By observing how his computer-drawn sperms move, Barr can figure out how fast, in what direction, and with what kind of style the sperms travel.

Theoretical "improved" sperm can be designed with a computer, be animated, and then be observed on a TV screen.

His research began as a study of the mechanism of cell movement. Since a sperm cell has one of the simpler mechanisms, he decided to focus on that. He's been trying to design the fastest head-tail design on his computer. "I'm analyzing the sperm as if it were a machine," Barr explains.

He has discovered that there is a very specific head-tail correlation. Sperms that power themselves with a corkscrew motion, for example, move fastest if they have a flat head, "like a pancake." Also, the head size of human sperm carrying the X chromosome, responsible for girl babies, is a little larger than that of the boy-producing Y sperm. But since boy sperms are slightly faster, he says, this might explain why statistically more of them are born each year.

Although his work is pure research, Barr says it may give biologists, fertility experts, and other reproductive researchers insights into improving their artificial-insemination techniques.

RED-TAPE MEASURES Next time you confront a bureaucrat, take a careful look. If the hair is thin, you can reckon he has a big budget behind him. A bushy pate means his department is penniless. A high forehead? The chances are that he directs droves of employees.

These unlikely images do not belong to real bureaucrats. Instead, they are *symbols* that come from the new science of bureaumetrics, the brainchild of social scientists Christopher Hood and Andrew Dunsire, who contend that bureaumetrics does for public administration what econometrics does for the financial world. Their system subjects every aspect of bureaucracy to intense scrutiny, analyzing every agency expenditure, every employee's workload, and even the qualifications of the specialists within a department.

Because it injects statistics into areas traditionally clogged with folklore, bureaumetrics can also be used to make historical analyses. The researchers recently discovered, for example, that during the *laissez-faire* era of Queen Victoria, when minimal government was the rule, the British civil ser-

> The blood-vessel pattern
> of the human eye is distinct
> and individual enough
> to serve as identification.

vice grew twice as fast as it would grow from 1920 to 1970, when government participation is said to have reached unprecedented proportions.

What does all this have to do with faces? Actually, the faces are computer drawings meant to help analysts characterize individual bureaucracies. Each facial feature, like a bar on a bar graph, is the measure of one critical quality, such as department size or efficiency. To glimpse his own image, an agency director can feed bureaumetric data into a computer equipped with a special program built by Dunsire and Hood.

All this the researchers have put into their new book, *Bureaumetrics,* which might become required reading for prime ministers and presidents keen to cut the flab out of their governments.

EYEDENTITY Someday you may be able to withdraw your money from an automatic teller machine by staring at it. This sort of eye contact is just one of the uses for the EyeDentifyer, an automatic identification machine invented by Robert B. Hill, of Portland, Oregon.

Hill had been working on a medical tool with his father, an ophthalmologist, when he realized that each person has a unique pattern of blood vessels on the back of the eye. To analyze these patterns, he built the EyeDentifyer, a small, black microprocessor-controlled box that contains a light source and an optical scanner. The subject looks into a window in the box, focuses on a "fixation target," and pushes a button. The EyeDentifyer automatically scans and records the subject's blood-vessel pattern, converting it into computer data that are compared with an eyeprint already stored in a bank or on a credit card. By matching the live eyeprint with the stored one, the EyeDentifyer quickly verifies identity.

Hill says that his device will usher in the age of automatic personal identification. The machine can verify checks or credit cards, secure electronic funds, and control access to high-security areas on military bases or in nuclear-power facilities, and it has already caught the eye of the CIA. Right now

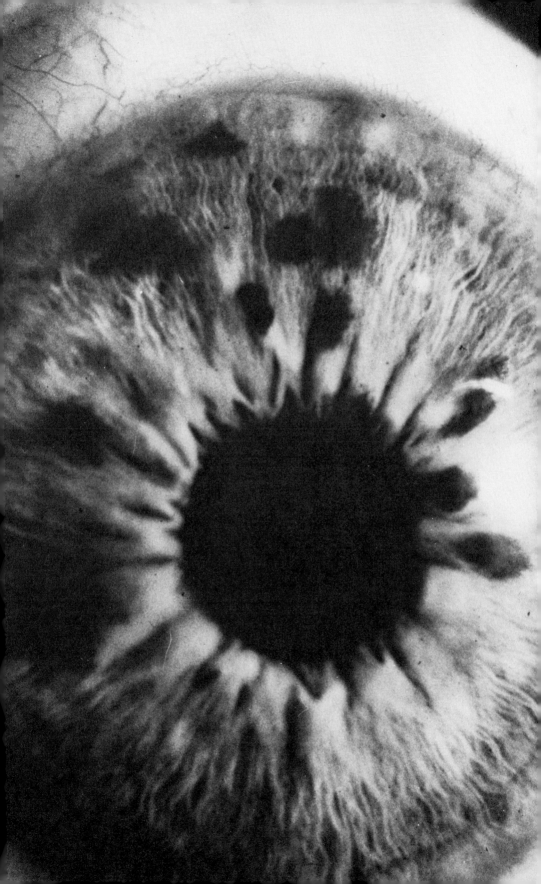

Cheap scotch—or the real thing? The new science of chemometrics can solve this most important of problems.

the EyeDentifyer is still a laboratory prototype, but Hill's company, EyeDentify, Inc., plans to market it soon.

ULTRASONIC STEAK The difference between "prime" and "choice" steak is a matter of personal preference. USDA meat inspectors use the subjective methods of sight and touch to determine the grade of beef on the hoof, and thus the grade of your steak.

This may change. Paul Gammell, of the California Institute of Technology, working for the Jet Propulsion Laboratory, in Pasadena, has developed a way to grade beef scientifically, using ultrasound.

Beef is a mixture of fleshy and fatty tissues. Muscle and fat reflect high-frequency sound waves differently, and so different mixtures will have different ultrasound reflectance patterns. The less fat in the tissue (that is, the better the grade of beef), the richer will be the ultrasound patterns.

Gammell used an ultrasound transducer to measure the ultrasonic patterns from 40 steaks, 10 from each USDA grade (prime, choice, good, and standard). He found definite correlations between grade and sound patterns and concluded that it is perfectly feasible to establish a scientific butchered beef grading system by using ultrasonic scanning.

It should even be possible, he adds, someday to grade beef cattle ultrasonically before they ever get to the slaughterhouse.

CHEMOMETRICS With an inspired use of analytical mathematics, one University of Washington chemistry professor has helped to found a new subspecies of chemistry and has, among other things, given forensic chemists a way to spot that staple of clip joints, cheap scotch in premium bottles.

Dr. Bruce Kowalski points out that one of the shortcomings of analytical chemistry is that it does not use very sophisticated math to analyze the data it collects. He's trying to correct that with a new branch of chemistry he helped establish called chemometrics. This joins the excellent data-gathering skills

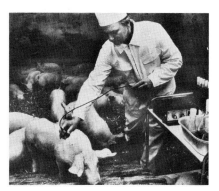

Researcher measures a pig with ultrasound as part of an experiment in improving meat-grading techniques.

of a good chemist with the analytical skills of a good statistician, for example.

Chemometrics is especially valuable outside the laboratory, where substances to be analyzed are impure. Dr. Kowalski helped the Wyoming State Crime Laboratory come up with a simple test to tell whether inexpensive booze, such as scotch, has been introduced into high-priced bottles.

Chemically the problem was tricky, since the chemistry of scotch changes if the bottle has been opened or if the liquor has been mixed with water. Kowalski's solution was to take different samples of premium and cheap scotch and first run them through a gas chromatograph, a device that sorts out molecules according to size.

The tests came up with hundreds of components, but, with the help of a computer recognition program that analyzed the results, it narrowed down the cheap scotch/good scotch factor to two chemicals that chemists could look for.

Chemometrics can yield detailed descriptions of a variety of substances, Dr. Kowalski notes. His tests can identify and isolate distinguishing factors in crude and fuel oil, for example, making it possible to decide who is responsible for an oil spill. Aerospace manufacturers have asked him to work up chemical profiles of special materials they need. And just recently Dr. Kowalski matched his technique against a panel of human wine tasters to find out just what it takes to make a superior Pinot Noir.

ELECTRIC ROADS And now . . . electric freeways.

Led by Carl Walter and Steve Wilson, a team of researchers at the Lawrence Livermore National Laboratory (LLNL), near San Francisco, has built a 220-meter-long powered roadway that provides energy to a 1969 Volkswagen.

Buried in the roadway is a magnetic steel core, its top flush with the surface of the road. Inside the core are six insulated aluminum cables, carrying from 500 to 1,000 amps of alternating current. An electrical pickup in the

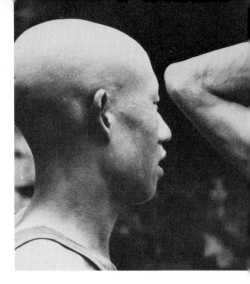

car's bottom collects the electricity from the cables by induction, without ever touching the surface, and turns the alternating current into direct current for the car and its batteries. The vehicle runs on battery power on "unpowered" roads.

This isn't the first time someone has thought of using inductive energy transfer; the first patents were issued in 1891. This modern version may find use in Santa Barbara, California, for a downtown electric bus system.

As for electric freeways: LLNL engineers estimate that running a roadway-powered electric car should cost between 1.2 cents and 2.4 cents per kilometer. The cost of turning regular freeways into powered roadways could be from $217,000 to $373,000 per kilometer.

LASERS VS. BALDNESS? If the purveyors of balding "cures" discover some recent Chinese research, the next useless thing applied to bald spots may be the laser beam.

According to the Chinese-language *Laser Journal*, scientists from Guangdong Province shaved hair from two areas on each of 12 guinea pigs. They illuminated one area on each animal with an infrared laser for a minute a day and left the other shaved area alone. After 20 days the researchers found that hair grew faster in the laser-treated areas: 0.71 millimeter per day on the average versus 0.59 millimeter per day in the untreated areas. In addition, there were more hair follicles and hair was "thicker, longer, and more uniform" in the laser-treated spots.

What works on guinea pigs with normal hair, however, won't necessarily work on men whose hair has stopped growing. The Chinese study did *not* mention human baldness. But a lack of scientific evidence probably won't prevent charlatans from marketing the laser "cure."

RECOMBINANT DIARRHEA Now there's yet another triumph for recombinant-DNA technology on the horizon: a cure for diarrhea.

Baldness treatment?
In China, scientists report
that lasers promote
hair growth on guinea pigs.

Dr. Stanley Falcow, head of the department of medical microbiology at Stanford University, in Palo Alto, California, has isolated the gene in a toxic form of the intestinal bacterium *E. coli,* which makes the poison that causes the runs. Dr. Falcow conducted his diarrhea research not at Stanford but at the University of Washington, where he teamed up with Walter Dallas, Magdelene So, and Steve Moseley.

The four scientists used recombinant-DNA techniques to neutralize the diarrhea gene and, in the process, made possible a vaccine. The Cetus Corporation, in California, a gene-splicing company, is currently conducting laboratory tests with the vaccine on pigs.

An embarrassing inconvenience for tourists, diarrhea is a major health hazard for children in poorer countries and a leading killer of livestock, particularly piglets and calves.

Because the *E. coli* genes in large farm animals and in humans are so similar, a vaccine for human diarrhea is quite feasible. Moseley, working with the World Health Organization, has already developed a simple test to pinpoint people harboring the toxic microbe.

The work by Falcow and his associates may also lead to a new cure for whooping cough and possibly cures for cholera, dysentery, and certain venereal diseases.

LONG-LIFE BANANA The idea of a long-life banana—or, for that matter, long-life apples, pears, and mangoes—might sound like something out of the Restaurant at the End of the Universe, but that is what will shortly be on supermarket shelves, thanks to an idea being developed by Tal Chemicals, a wholly owned subsidiary of the British sugar giant Tate and Lyle.

The secret is not some sort of plastic coating—that has been tried in the past with little success—but a sugar-based solution into which the fruit can simply be dunked and allowed to dry.

The result should be sweeter, better-quality bananas, with a long shelf life

" 'Not quite proved'
in mathematics
is like 'not quite
pregnant' in biology."
                    —Howard Pattee

and the saving of millions of dollars in refrigerated ships used to ferry around the world the 7 million tons or more of the fruit consumed annually.

Originally developed by Peter Lowings, of the Department of Applied Biology at Cambridge University, the coating, known as Prolong, slows down the movement of gases such as oxygen and carbon dioxide across the skin of the fruit. This slows respiration, effectively delaying ripening.

At present, bananas are picked when they are immature and are shipped at temperatures around 13.5°C to delay ripening until they arrive at their destination. When Prolong was used in the Caribbean, fruit was shipped successfully at 21°C, without any increase in the number that became overripe on board; shipments from the Philippines to Hong Kong at temperatures of up to 32°C also arrived in good condition.

According to Nigel Banks, one of the Cambridge scientists, the reduced need for refrigeration might mean an $8 million savings on the capital cost of each new banana vessel, on top of the savings in running costs.

The coating, which is principally long-chain lipids with polysaccharide molecules, has also been shown to be effective for apples, pears, plums, avocados, and mangoes. It has little effect, however, on grapes, tomatoes, and strawberries. These lack the stomata that the coating is thought to block.

Prolong is already in commercial use on apples and pears in England and on some bananas shipped from the Philippines and the Caribbean, although, as far as the United States is concerned, the coating, which is essentially a natural product, is still awaiting Food and Drug Administration approval.

Banks says, "Being cheap, nontoxic, water soluble, and easy to apply, Prolong may well find widespread use where more complex postharvest treatments are not feasible." Work is under way to tailor the product to other fruits.

Tal Chemicals says that those worried about a coating of sugar around their apples, pears, and other fruit need have no such fears. "It works out at

Grocery shoppers may not realize it, but the variety of vegetables for sale results from a kind of "genetic engineering."

about one crystal of sugar per fruit," the company said. "That's not going to give anyone galloping tooth decay or an expanding waistline."

VEGETABLE EVOLUTION No one really knows when man first began to mimic nature by selecting the best plant from his fields and mating it with another to produce a better crop, but the process has continued ever since. The result has been a huge assortment of varied vegetables and fruits at the supermarket.

The potato, which orginally came from Peru, was only about the size of a prune. Farmers later bred it for size.

Corn resulted from some wild grass through selective breeding by the ancient Indians of Central America. The grass had only six or seven kernels on it. Modern corn has thousands and is so highly domesticated that it has become what scientists call a biological monstrosity. It can't survive in the wild because it has no way of spreading its seed. It can produce offspring only if farmers remove the kernels from the cob and plant them themselves. But from this small grass has come many supermarket varieties: red corn, yellow corn, sweet corn, flour corn, and popcorn.

Just parts of some plants are selected. For example, some people selected the leaves, some the roots, and some the sugar of the beet. As a result, the garden beet, the sugar beet, and the leafy green Swiss chard are all derivative forms of that one beet species.

This also applies to the mustard species, *Brassica oleracea*. Through the centuries it's been modified in many ways. From the terminal bud has come the cabbage; from the flowers, cauliflower; from the lateral buds, Brussels sprouts; from the stem and flowers, broccoli; and from the leaves, kale. All evolved from artificial selection from the wild. And as different as they may appear, they are all from one plant.

With recombinant DNA research now on the rise, newer and more unusual plants from the original will soon be appearing on the supermarket shelves.

Cyrus Vance:
His comings and goings
were all fed
into a Yale computer.

SILICON CYRUS Cyrus Vance, the Secretary of State under President Carter, may be forgotten in the minds of most Americans, but he's still a topic of conversation at Yale's computer science department. The Yale Artificial Intelligence project there has programmed Vance's memory into one of their computer programs.

In reality, the program, called CYRUS, was designed to see how human memory works without actually poking around inside the brain. Vance was just available for the research—or rather, his comings and goings were. The scientists developed the "smart" program by gathering up all of the news stories about the ex-Secretary from the news wires and then fed the info into the computer. What they ended up with was Vance's memory in silicon.

Smart programs are a quantum leap from present-day programs. "Most memories," explains Roger Schank, the head of the project, "have static, pre-assigned categories that don't or can't change, and more important, have no relationship to the meaning of what's in them. In other words, they're just sets of boxes within boxes."

The smart programs, like our brains, create categories as they receive information and then change the categories when some new information is injected into them. "It's a sort of self-adapting dynamic memory," adds Schank. Every new bit of data causes CYRUS to reorganize everything it knows. Unlike other computers, this keeps every piece of information in proper relation to every other. The "smart" programs, Schank believes, mimic human thought. If the information isn't available, the computer will make its own conclusions from what it does know. Other computers can't do this yet.

As of late, Schank and his associates have gone beyond CYRUS and into something even more complicated. "We have a program," says the computer expert, "that is becoming an expert on terrorism. It's reading all of the terrorist stories until it knows more about the subject than anyone else." The project is funded by the Department of Defense.

*"Man is still the best computer that
we can put aboard a
spacecraft— and the only one that
can be mass-produced with
unskilled labor."*
—Wernher von Braun

SUPER-PRESSURE SUPERCONDUCTORS Computers made of super-conducting materials that transmit electricity at high speed would work hundreds of times faster than the computers around today. Unfortunately, until now scientists have been able to make superconductors only by chilling certain nonmetallic materials to temperatures just above absolute zero. At such frigid temperatures, the molecular structure of these materials changes, enabling them to transmit electric current with absolutely no electrical resistance.

But a barrier to this "supercooling" has always stood in the way: The price of the cooling process is monumental. Most research to solve the cost problem has been aimed at finding superconductors that could function at room temperature.

Recently, scientists have done just that. They have learned that if they subject potential superconductors to thousands of pounds per square inch of pressure, the molecular structure will change much as it does by cooling. A team at the Benet Weapons Laboratory, in Watervliet, New York, has already exposed the mineral greenockite to 40,000 pounds per square inch of pressure and produced a superconductor that works just under room temperature. And the Russians say they have converted copper chloride into a superconductor that works *at* room temperature.

Other scientists are using superpressure to create what could be the most powerful superconductor of all—a substance called metallic hydrogen. To make metallic hydrogen, scientists must subject hydrogen gas to a pressure of 15 million pounds per square inch. Because this huge pressure can be exerted on only a fraction of an inch of surface area at a time, however, researchers have been able to produce just a few molecules of metallic hydrogen. The effort to produce large quantities of the substance continues at six major American labs.

Meanwhile, according to engineer David Kendall of the Benet Lab, researchers are trying to find materials that can be turned into superconduc-

Robot assembly line:
The Air Force plans to have a
robot factory that will
build airplanes from start to finish.

tors at lower pressures. Some of these substances could reach the market in about five years.

AIR FORCE ROBOTS Guided by their television camera eyes and electronic skin, the robot workers assemble metal sheets into wings, tails, and bodies. These airplane parts then travel on to another chamber, where more robots finish building the plane.

This scenario should be reality by 1985, when the United States Air Force completes its Factory of the Future, a robot-operated factory that builds airplanes from beginning to end. The Future Factory, says the air force, will be four times as productive as the factory of today.

According to Michael Moscynski, manager of robot efforts at the Air Force Materials Laboratory, in Dayton, Ohio, the new factory will be controlled by a supercomputer that coordinates thousands of smaller computers and hundreds of robots. These worker robots will be divided into groups of three or four, with each group at work in a factory "cell" dedicated to a particular task. Airplane parts will move from cell to cell so that the different groups of robots can drill holes or rivet metal. In a few days, a complete airplane will emerge.

The air force is working with General Dynamics, Lockheed, and McDonnell Douglas, aerospace companies that hope to have Future Factories within the decade. All three are testing intelligent robots that can "see" and "feel" their way around specially designed cells. McDonnell Douglas is even developing an electronic language that will enable robots to speak with their fellow robots around the factory floor.

STEALTH PLANE It's no secret that the Pentagon has spent years trying to produce planes that cannot be detected by radar. Thus, aviation experts were hardly shocked when word leaked out that the United States had developed a "stealth" aircraft capable of eluding Soviet radar almost completely.

Radar outposts:
Stealth aircraft would be
able to elude
all such early-warning devices.

Two things, said former Defense Secretary Harold Brown, make the craft unique: its shape is designed to reflect the fewest possible radar waves, and its special coating diffuses radar beams. While the Pentagon is vague about other more specific design details, anyone who reads technical journals can figure out roughly how the stealth plane would look. Tiny and slickly streamlined, it would present the smallest possible surface to radar waves. It would be made of graphite, epoxy, resins, plastics, and asbestos—materials that absorb and diffuse radar. And it would incorporate special materials to insulate hot parts of the engine, thus shielding them from infrared or heat-seeking devices.

Rumor has it that at least one of these planes crashed on a test flight, possibly because of its unusual shape. Nevertheless, Brown called the stealth "a major technological advance" that "alters the military balance significantly."

CHUNNEL UPDATE The *Gossamer Albatross* human-powered airplane may have made it over the English Channel, but engineers have had less luck with their plans to go under the watery separation between England and France. Back in 1975 plans for an ambitious tunnel beneath the channel— the Chunnel—were scrapped because of technical and financial problems. Now Tarmac. Ltd., an important British construction consortium, has submitted new plans for the Chunnel, and the project may be alive once more.

A revision of the original plan, Tarmac's proposal calls for construction of two rail tunnels with a connecting supply and support tunnel between them. Completion would come in the year 2000, with one rail tunnel ready in 1990 so that limited service could begin sooner. Terminals would be at Cheriton, near Folkestone, in England and at Frethun, near Calais, in France.

The cost of the project has been estimated at $5.2 billion, half of which would come from England and half from outside investors. Construction would begin in 1984.

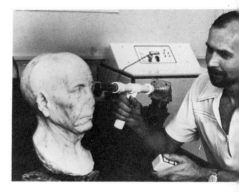

Vaporizing dirt
and grime from a statuary
is just one of many
new jobs for the laser.

Tarmac produced the report with the help of a small team of in-house experts, who determined that interest in the Chunnel was still high even though the British government had announced that no government money for the project was available.

Eric Fountains, Tarmac's chief executive, states that: "Discussions already held with various organizations in France and elsewhere on the Continent confirmed their interest and underlying keenness to proceed."

AMAZING LASER FACTS If you've ever seen a laser, it's probably been a box a little over a foot long that emits a red beam. Such helium-neon lasers have become common in places ranging from high-school physics labs to supermarket checkout counters, where they read the cryptic striped codes on food packages. In the world of science such things are considered mundane because of some of the other things lasers can do:

• Periodic motion only about as large as the diameter of a proton—10^{-15} meter—has been measured with a laser.

• A huge laser at the Lawrence Livermore National Laboratory can, for a billionth of a second, deliver more power to a 0.01-inch pellet of hydrogen than the country's entire electric power system can generate. The result is thermonuclear fusion in a "microexplosion" similar to that of a hydrogen bomb but on a scale so much smaller that it can be confined in a special chamber for the study of fusion physics or the effects of nuclear weapons.

• The beam from a helium-neon laser can be used in the place of acupuncture needles. The power can be so low that the patient doesn't feel any warmth from the light.

• By shining a laser beam through the lens of the eye, a physician can weld a detached retina back to the eyeball without cutting into the eye.

• Lasers drill 0.01-inch holes in cigarette paper to help ensure uniform air flow and thereby control the amount of tar in the smoke.

• Lasers have detected air pollutants in concentrations as small as a few

The blast furnace is quickly being outmoded by new technologies producing ceramics and newfangled polymers.

parts per billion and have been able to detect single atoms in the laboratory.

• Laser light has been used to *cool* atoms to within a fraction of a degree of absolute zero (-273°C or -460°F), the temperature at which, in theory, the constant motion of atoms comes to a complete halt.

MIRACLE MATERIALS "The 1980s," says George B. Kenney, of the Massachusetts Institute of Technology's Materials Processing Center, "are the decade of materials. You're going to see a lot of changes in both the materials that we use and the way that we make them."

Once wood was wood, metal was metal, and glass was glass. No longer. Now, using microprocessing and high-tech manufacturing methods, engineers are crafting a whole new breed of materials for space and industry.

These include ceramics that can be stamped out in metal-bending presses, metal powders that can be mixed into once-impossible alloys, enzyme-generated plastics (made without fossil fuel), and reinforced polymer resins that are lighter than aluminum and stronger than steel.

Metals producers are experimenting with supercooling: chilling molten metal so fast that it doesn't have time to form typical crystals. It comes out as a kind of metallic glass or as a fine powder that is stronger when reformed than the original metal ever was.

Newfangled polymers will make up our cars and planes in a few years. "What aluminum was to the 1920s," predicts an industry official, polymers will be to the 1980s. "By the 1990s," says another, "an airplane's major components will be graphite."

Materials experimentation is even edging toward genetic engineering. By molding life processes and the chemical composition of building materials, scientists may be able to produce a steel or a plastic that grows itself.

JAPAN'S SUPERCOMPUTER The Japanese have a deep desire to prove that they can do more than improve on other people's technological break-

throughs. While their economic clout has steadily grown and their reputation for engineering excellence has expanded, the Japanese have still not entered the big leagues of ground-floor technological development.

Now our Eastern competitors want to change all that by blitzing the computer world. They intend to create a fifth generation of computers that goes as far beyond today's machines as the latter are beyond UNIVAC. Basically, the Japanese have announced that they will attack and solve all the remaining problems of today's computers, then build a machine that incorporates all the solutions.

The Japanese machine will be able to listen to and talk in a facsimile of human language. It will store information in a manner much more like human memory. It will probably use superconducting Josephson Junction switches, so it will be at least 100 times as fast as today's best machine, possibly 1,000 times as fast.

Japan's International Trade Ministry has made development of the fifth-generation computer by 1990 a national goal. Some $50 million has been earmarked for the first three years, with much more to come later. Reports of a $450 million commitment have swept the computer industry. A Japanese official has said that $1 billion is possible. The Japanese government is jaw-boning companies to cooperate in a single unified effort to create the new machine.

Considering the government's well-known ability to throw the entire nation behind a stated goal, American computer makers are nervous. They have decided, at the very least, not to help the Japanese development in any way. In late 1981, Japan held an international conference on the fifth generation. The American computer community en masse refused to attend.

MOVING BUILDING The world's most unique office building is putting itself together in Florida. The new headquarters of the Gilman Paper Company will have a framework of beams put up by normal construction methods, but

This is one way of moving a building. But the Gilman Paper Company has a more modern approach.

beyond that the building will construct itself, in consultation with Gilman employees.

Experimental architect Cedric Price has devised a system in which workers can develop, with a computer, design requirements for their own working areas. The computer then prints out instructions for a crane operator, who moves cubes, walls, and other building sections around to suit the workers' needs.

The building can literally change its shape overnight. If the workers need a temporary meeting room, they can have it. If a new manager wants a different office setup, he can get it. A team of architects created the computer software that allows the computer to redesign the building without overstepping structural bounds.

Gilman says his building is "an aid to the extension of one's own interests," that will give workers an opportunity "to work, create, think, and stare." The thinking office's framework is in place and the 150 four-meter cubes that will form the core of the jigsaw interior are being built. It should open sometime in 1982.

CHROME INDEPENDENCE Without chrome, we couldn't build aircraft engines or power plants. Chrome, when added to stainless steel, is the only metal whose atoms bind to the oxygen atoms, forming an impervious skin that protects any object exposed to extreme temperatures or chemical attack.

But because most chrome comes from South Africa and Zimbabwe-Rhodesia, our supply is in grave danger. In the near future, the two countries could easily form an OPEC-like cartel, severely raising chrome's price and limiting its worldwide distribution. Even worse, a war in the African continent could cut off our supply of the metal altogether.

According to Elihu F. Bradley, a materials expert for the Pratt & Whitney Aircraft Group, of Hartford, Connecticut, a shortage of chrome would be as

Assaulted computers:
In greater and greater numbers,
people who don't
like computers are striking back.

devastating to the United States as an oil embargo. "Minerals independence should be a national goal," he says, "just as energy independence is."

Bradley suggests that the government deal with the threat by forming a department of strategic metals modeled after the former Department of Energy. The office, says Bradley, would devote itself to finding substitutes for chrome, as well as for other critical imported metals, including cobalt, platinum, and manganese.

"It's really hard to say which elements or alloy systems might make the best substitutes," Bradley notes. "But this sort of effort would probably cut the nation's appetite for chrome and cobalt in half by 1990."

MURDERED COMPUTERS There are two kinds of people in the world— those who understand computers and those who don't. And the "don'ts" are beginning to fight back—with a vengeance.

Gary W. Dickson, a computer expert at the University of Minnesota, has documented numerous cases of computer murder throughout the United States. For example, a sheriff in California shot a computer dead, for uncontrollably spewing out arrest records. Someone at the Minneapolis Post Office poured honey into newly installed terminals. Car keys were deliberately tossed into a disc file in Denver. A computer at the Metropolitan Life Insurance headquarters, in New York City, was brutally attacked with a screwdriver.

These are only the violent assaults. More commonly, says Dickson, employees sabotage the machines by intentionally inputting mistakes into the system. Managers engage in their own form of abuse by blatantly ignoring computerized information in their decision making or overlooking computer advances.

Dickson, director of the University of Minnesota's Management Information Systems Research Center (MIS), says, "Use of the technology is not nearly as sophisticated as the technology itself."

*"The two big tricks of the twentieth
century are: technology
instead of grace, and information
instead of virtue."*

—Ulysses Cantois

MIS, along with a similar program at the Massachusetts Institute of Technology, specializes in training managers to deal effectively with the stresses and strains placed on a corporation's employees when a computer system is being implemented.

"An information system tampers with the nervous system of the organization, its power structure," Dickson explains. "The company's got to involve the ultimate users, the employees, in the design process. And handholding is necessary. You can't just plug a system in and walk away."

Or, at least, not if you don't want honey in your terminals.

HYDROGEN FACTORIES If the hydrogen on Earth were readily accessible, it could become a vital and unlimited source of energy. But most terrestrial hydrogen is locked up, bound with oxygen to form water (H_2O).

For hundreds of millions of years, only the cells of the green plant have been able to split water into hydrogen and oxygen; they have accomplished this through the use of chloroplasts—tiny, disc-shaped structures that manufacture carbohydrates.

Now man is copying nature. Melvin Calvin, a chemist at the University of California at Berkeley, has produced synthetic chloroplasts. Calvin believes that once we put these artificial chloroplasts to work on splitting water, we will have access to an infinite amount of hydrogen fuel.

Natural chloroplasts split water whenever they are hit by sunlight: A ray of sun knocks an electron from a chloroplast, which compensates by sucking an electron from water. Without its electron, the water grows so unstable that it splits into hydrogen and oxygen. The hydrogen then combines with carbon dioxide from the air to form the carbohydrates needed for the plant's growth.

Calvin's artificial chloroplasts function in much the same way, except they are not designed to help hydrogen combine with carbon dioxide to form new plant material. Instead, they release the hydrogen as a gas that can power cars, and provide fuel for homes and factories.

Any shutterbug who's taken a tricky picture has practiced photochemistry. Now there are some striking new applications.

By the turn of the century, Calvin says, we're likely to see synthetic chloroplast factories everywhere.

LASER PHOTOCHEMISTRY In summertime beach bums like to saunter off to the sand, lie in the sun, and practice a little photochemistry. All those tourists with cameras slung around their necks are photochemistry devotees, too.

Photochemistry, the use of light to create or augment chemical reactions, has been a peculiarity of chemistry for decades. Suntanning is photochemistry; so is the photographic process. But for all its pervasiveness, photochemistry has been a bust as a general-purpose scientific technique.

Now the growing use of lasers has created new interest in photochemistry outside the camera and darkroom. The process is being considered for separating rare isotopes, creating more efficient semiconductors, and synthesizing specialty chemicals.

The problem with photochemistry has always been that because light contains a wriggling mass of frequencies, of which the entire rainbow of colors is merely a snip, it is chaotic and hard to control. So, creating reliable reactions with light has always been difficult. The laser, however, eliminates this problem, because it is pure and constant light. Even better, the laser can be pinpoint-focused for meticulous reagent processes.

Experiments at Los Alamos National Laboratory, for example, show that photochemistry has the potential for creating a new manufacturing process for a rare uranium isotope needed to make nuclear fuel. In the reaction the laser neatly slices off a fluorine atom from uranium-flourine gas, leaving the nuclear material behind.

In the future, researchers see lasers making possible photochemical reactions in which chemists, like brain surgeons, can move atoms around at will, pushing them with light energy, crafting new chemicals whose existence is impossible today.

Shocks of corn stalks:
The leftover husks and stems
can now be
turned into a plastic.

METALLIC GLASS Scientists at Allied Corporation have created a material so strange one executive calls it "a new type of matter." It is as strong as steel, can be bent like metal, and doesn't corrode. It is metallic glass, and it may save America's utilities billions of dollars in wasted power.

The metallic glass results from processing the original materials at precisely controlled levels of extreme temperature and pressure only recently possible. The result is a material with the structural properties of metal but the inertness of glass. It can be made much more cheaply and quickly than any conventional metal.

The biggest impact of metallic glass will be on electrical motors and transformers. The metal parts of today's electrical transformers soak up 2.2 percent of America's power production, more than $1 billion a year worth of electricity. If metallic glass were used in the transformers it would cut this loss by 60 percent, saving utilities some $600 million a year and increasing American energy output by 1 percent without a single new power plant.

CORNY PLASTIC Until recently the only economically worthwhile thing midwestern farmers could make from their leftover corn husks and stems was gasohol.

Now a new possibility has arisen: turn the corn into plastic. The key to the process is enzymatic transformation, a new type of chemical process that substitutes specific enzymes for the heat and pressure of traditional chemical methods. Enzymes are body chemicals that shuttle organic molecules around, linking them into new chains, creating new chemicals from old. By exposing a material to enzymes in a prescribed order, scientists can, step by step, transform one organic material into another.

In the case of corn, the nation's second-largest corn syrup refiner, A.E. Staley, of Illinois, has set up a production facility to turn corn into a plastic intermediate. The plastic product made there will aid in the production of insulation, packaging films, and alkyd resins.

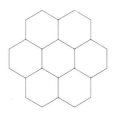

PHENOMENA

CHAPTER 7

The first photograph
of *Rheobatrachus silus* mother
giving birth to
its tadpoles through the mouth.

MOUTH BIRTH With a little patience and a unique amphibian, two Australian researchers have managed to get the first photographs of an unusual style of birth: through the mother's mouth.

A squat, pug-nosed, dull-colored frog from Australia, called *Rheobatrachus silus,* incubates its young in a strange way. The female swallows fertilized eggs, and they develop into froglets in her stomach. When the froglets are ready to emerge, the mother frog vomits them up to her mouth and lets them hop out and away.

Michael Tyler and David Carter of the department of zoology at the University of Adelaide, Australia, were able to photograph both natural and induced births in two of these frogs. One mother shot out six of her young in less than one second. They landed over a foot and a half away. Those that didn't hop away, the mother swallowed again. Taking pre- and postbirth weight, the scientists found that the froglets accounted for over two-thirds of the mother's weight.

According to biologist Stanley Salthe, of Brooklyn, New York, these frogs can perform this unique brooding because of their low metabolism. They can get by with little oxygen, breathing through their skin rather than their lungs, which are crowded by the bulk of the small frogs. And since eating would stir up gastric juices that might destroy the eggs, the frogs can also go without eating for a long time.

BOOK DEATH To the horror of librarians across the country, many of the books printed in the mid-nineteenth century are turning to dust. At Stanford University, for example, 27 percent of the 1.5-million-volume humanities and social science collection is too brittle to use.

The main reason for the books' deterioration, says Paul Banks, a professor at the Columbia University School of Library Service in New York, are the acidic and weak materials that were introduced by nineteenth-century paper manufacturers. According to Banks, alum-rosin, the substance that prevents

While libraries may appear massive and permanent, many of the books they protect are crumbling to dust.

ink from spreading on a page, combines with sulfur dioxide and moisture in air to form sulfuric acid, which decomposes paper.

Another problem, says Banks, is the substitution of wood pulp for rags. While the long fibers in rag papers have helped keep fifteenth-century books in excellent condition, the short fibers in wood pulp are easily weakened and broken. As paper is still being made with wood pulp and alum-rosin, Banks expects today's cheap paperbacks to crumble by 2020.

The enormous expense of repairing the brittle books has led libraries to other choices, such as microfilming and photocopying on nonacidic paper. Libraries also use air-filtering systems and temperature and humidity controls to slow the books' decomposition.

But for some brittle books, the preservation effort will come too late. According to Sally Buchanon, conservation officer at Stanford University Library, "There are going to be huge gaps in research collections in the future in materials from the nineteenth and twentieth centuries."

GREEN HAIR Shortly after moving to Columbia, Maryland, Peggy Fenzel, a natural blonde, noticed something odd. Her hair was changing color. Weeks later her friends confirmed her suspicion: She had green hair.

So did other adults and children with white or light hair in the neighborhood where she lived. In fact, some children, like the main character in the movie *The Boy with Green Hair,* were having hard times in school because their classmates made fun of them. "I thought I was going crazy," Mrs. Fenzel said.

What began as a strange and annoying problem—"I couldn't wash out the green"—became more serious once the source of the problem was uncovered. Mrs. Fenzel's housing development is serviced by a private water company that draws its supplies from a large underground well. It turns out that the water is highly acidic, so much so that it dissolves minute quantities of copper from the plumbing in the houses.

167

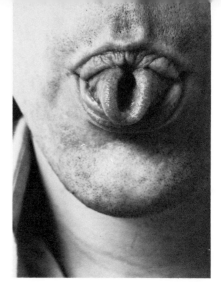

If you can roll your tongue into a lengthwise tube, you may be suited for a career in the life sciences.

After weeks of drinking and washing with the copper-rich water, Mrs. Fenzel and others absorbed some of the metal into their systems. And copper, when ingested by some people, has the peculiar property of turning hair color to green.

The copper discovery has given the Fenzels and other families more than green pates to worry about. Now Mrs. Fenzel says she's concerned about the effect of large amounts of copper on her and her family. Even before the copper discovery, when her baby was eight months old, she had stopped using the water to mix formulas for the child, who was always getting sick. The family has now switched to bottled water for drinking and cooking and uses the well water only for washing. In the meantime, Mrs. Fenzel and her husband have sued the water company and the Maryland Health Department to make the water clean, not green.

TONGUE TWISTER Can you roll your tongue into a lengthwise tube? Whether you can or can't is a genetically determined trait—a fact known for some years. However, researchers in the genetics department of Swansea University, in Wales, have also discovered that people who have this ability are likely to have an accompanying characteristic: the kind of personality that suits them to the study of the life sciences.

The Swansea geneticists found that 80 percent of life scientists surveyed could roll their tongues, while only 65 percent of the school's art students had this genetic skill. They concluded, therefore, that there is a relationship between the genes that encourage people to become scientists and those that control tongue rolling.

A quick survey through the *Omni* offices revealed that 72 percent of the editorial staff can perform the tongue-rolling trick—a score almost exactly midway between those of Swansea's scientists and its art students. Obviously, this is proof positive that *Omni*'s *Continuum* is a perfect blend of art and science.

Humanlike carrot: While this hand-shaped carrot is just an oddity, Swedish scientists say they have the real thing.

HUMAN CARROT Scientists in Lund, Sweden, have accomplished a remarkable feat: Loosely speaking, they've crossed a human being with a carrot.

To be precise, researchers at the Institute of Molecular Cytogenetics have fused human cells with carrot cells. "Many people think this is a dramatic experiment," says the institute's Professor Antonio Lima de Faria, "but the differences between humans and plants aren't as great as generally believed."

Using special enzymes, the research team was able to remove the carrot's thick cell walls in order to achieve the carrot–human cell fusion. The Lund scientists were also able to merge human cells with daisy cells.

The carrot-human fusion was part of the institute's basic research into chromosome organization and also helps to elucidate the problems of cell differentiations. Since cancer is essentially a disturbance of the cell-differentiation process, these experiments might prove valuable in combating the dread destroyer.

Dr. Lima de Faria, in reporting on his work in *Fauna och Flora*, the magazine of the Swedish Museum of Natural History, points out that animals and plants have many similarities: "Chromosomes in plants and animals have the same basic structure. What's more, although plants have no breasts, they still have the sex hormones found in humans. Pharmaceutical companies extract the sex hormones used in birth-control pills from plants.

"Furthermore, the urine of pregnant women," the professor writes, "contains estrone and estriol, which are also found in palm trees and willows."

URBAN LEGENDS In ancient times they had the *Aeneid*. In the Middle Ages it was the Holy Grail and the Knights of the Round Table. And in modern times it's dead mice in Coke bottles and french-fried breaded rat.

That's the stuff of which modern urban legends are made, says University of Minnesota sociologist Gary Fine. Over the past few years he's been col-

By screaming out all
our stored-up pain, say two
researchers, we
can actually grow taller.

lecting and investigating the truth of such stories.

Many of these, he says, tend to revolve around fast-food chains and food conglomerates. "With our fear that impersonal corporations are changing our society and will do anything for a buck, the conditions are ripe for fear and alienation," explains Fine. And some of the stranger urban legends are symptoms of that fear, such as the rumors about spider eggs being found in Bubble Yum or those about McDonald's wormburgers—worms being used as filler in the fast-food burgers. These rumors are untrue, according to Fine, who painstakingly checked out every story. (The wormburger legend was promptly dismissed when one hamburger-store owner pointed out that a pound of worms costs more than a pound of ground beef.)

A check of appeals-courts records, however, did turn up the fact that dead mice, says Fine, *have* been found in Coke bottles as well as in soda-pop bottles of other brands. He has documented cases, from the years 1914 to 1976, of 45 dead mice found crammed inside bottles. Nearly all the appeals cases were against Coca-Cola, mainly because sales of that particular soft drink are so widespread.

Urban legends, explains Fine, orginate with those who use a product, are reported in the news media—usually as a denial—and from there spread like wildfire, with everything but the denial remembered by the public. And by processes he calls leveling and sharpening, only the gruesome points of the case survive.

Fine is now in search of his own Holy Grail of urban legends, french-fried rat. The story? A woman eats it while watching TV in a dark room. The facts? "Still checking," says Fine.

PRIMAL GROWTH When self-help devotees speak of "personal growth," they don't ordinarily mean it literally. But . . .

People who complete primal therapy (a therapeutic method known for its primal scream) may, in fact, actually *grow,* say the technique's inventor,

What do pilots do
just before their planes crash?
It seems that many
of them, believe it or not, whistle.

Arthur Janov, and his collaborator, E. Michael Holden, a neurologist.

Adults may grow 2.5 to 5 centimeters taller; men may sprout full beards and chest hair; and women may develop curves where there hadn't been any. The reason? Removal of stored pain.

The primal therapy patient undertakes a kind of inner voyage to the sources of embedded pain—in childhood and infancy, even *in utero*. (A few patients have "relived" their mothers' abortion attempts.)

About 20 percent of patients exhibit some kind of belated physical maturation, Holden says. Reliving and exorcising buried pain theoretically releases their full genetic potential.

"We have a model for this in psychosocial dwarfism," Holden notes, referring to well-documented cases of neglected or abused children whose growth was severely retarded.

The brain's limbic system, he says, acts as a sort of "capacitor." The stored pain is responsible for a neurotic, usually hypermetabolic physical state—abnormal heart rate, hormone levels, skin temperature, blood pressure, and other disturbances in vital signs. It is as if the body were battling a disease, says Holden.

WHISTLING AIR CRASHES Would you worry if you heard the pilot or crew members of an aircraft whistling during a flight? Some aviation-accident investigators think that perhaps you should.

Debate on the question began in the mid-1970s when Robert Rudich, an air-transportation consultant to the Federal Aviation Administration (FAA), revealed that "of more than two hundred sixty cockpit voice-recorder tapes removed from aircraft involved in accidents ranging from the minor to the catastrophic since 1966, over eighty percent have a recording of one of the pilots whistling during the last half hour of the flight."

Rudich attributed the whistling to "a sense of complacency" that led pilots or crew members to make "illogical" errors. Other experts, such as Gerard

This mermaid is made of sand, but those sighted by ancient sailors were probably images distorted by the atmosphere.

Bruggink, former director of the Society of Air Safety Investigators, questioned this interpretation, saying whistling can also have stressful connotations, including an attempt "to create an atmosphere of confidence in conditions of uncertainty or fear."

Paul Turner, of the National Transportation Safety Board, who went through the tapes to count the number of whistlers, and himself a pilot, cautions against worrying too much about it.

"Many times," Turner said, "things are always moving audio-wise in the cockpit. Your ears are always active. Then, when things get quiet, you whistle. I guess pilots just don't like quiet."

Turner admits, however, that "whistling is a reaction to stress in some cases," noting one incident in which pilots puckered up while a fire raged in the rear of the plane.

One researcher noted that the debate raises the important question as to whether pilots have too much to do at some times and too little at others, either of which might adversely affect performance.

The question cannot be dismissed lightly in any event. Rudich, who first raised the issue in a speech to the Lawyer Pilots Bar Association, ended his presentation by playing a composite tape from several aircraft accidents. Rudich noted that the tune one captain was whistling just before his Boeing 707 crashed was the first few bars of "The Battle Hymn of the Republic."

MERMEN AND MERMAIDS When Waldemar Lehn saw an armless figure with a large, bulging head rise above the surface of Lake Winnipeg, in Canada, he knew exactly what he was looking at. It was a merman, and he took a photo to prove it.

Lehn's picture, in combination with a clever computer program, showed that ancient sailors who reported mermaids and mermen were telling the truth. The men did see something: ordinary marine creatures whose images were freakishly distorted by changes in the atmosphere.

"In theory one is aware that the earth revolves, but in practice one does not perceive it, the ground upon which one treads seems not to move, and one can live undisturbed. So it is with time in one's life."
—Marcel Proust

Lehn, a professor of electrical engineering at the University of Manitoba, developed a computer program to simulate image distortions under a variety of weather conditions. From his readings in ancient Norse literature, he knew that mermaids and mermen were usually seen just before storms; thus, he concluded that common sea mammals, seen in a storm, would look nearly human, fitting the medieval description of half-human and half-fish creatures.

To prove it, he duplicated stormy-weather images of a walrus and a killer whale on the computer. The result: mermen shapes.

The cause of the distortion, according to Lehn, is a temperature inversion that occurs when a mass of warm air moves over cold air. This bends the light, so that objects are distorted beyond the horizon. As the inversion disintegrates, so does the image, becoming fuzzy or hairy-looking.

As for the mermen of Lake Winnipeg, Lehn finally learned that it was a foot-high boulder on the shore of the lake.

ANCIENT RECORDINGS Sounds that are thousands of years old, perhaps even human voices, may be recorded in the grooves of ancient pottery and glass artifacts, a Toronto pediatrician and amateur archaeologist now believes.

Dr. Peter Lewin, recognized for his work on ancient mummies, is trying to replay the chance sounds he thinks may have been captured on pots, vases, and plates from former civilizations. "If someone shouted or a dog barked close by as a vessel was being made on a potter's wheel, the sound could very well have been picked up on the vessel," Dr. Lewin says.

At first he attempted to "play" the pottery on a record turntable, using a diamond stylus. "That was too noisy," Dr. Lewin says. "So now we're using a laser similar to those on videodisc machines. The laser can be tuned to play only certain frequencies, such as those of the human voice. We've run three or four items through so far, without any success. Our next step is to try

Ancient phonograph records?
One scientist says all kinds of
sounds, even human voices,
may have been captured in pottery.

to prove a point. We're going to engrave a copper plate and see, experimentally, whether sounds can be captured in this way."

Although researchers in Europe are also using lasers in an attempt to recapture the fragments of sound that may be locked in the grooves of ancient artifacts, Dr. Lewin admits that the whole idea sounds like science fiction. "Some of my friends think I'm . . . well, put it in quotes, 'potty.' "

FUTURE FACES A few million years from now people will look much as they have been visualized by science-fiction writers: sleek, diminutive, toothless, and hairless, a Syracuse orthodontist claims.

Dr. David Marshall, who has studied the human skull for 35 years and who established an anatomical museum in Syracuse, New York, says the soft, refined foods that we eat are causing changes that will make our descendants somewhat odd-looking. "Man just doesn't use the jaw the way he did when teeth were weapons and food was unprocessed," he says.

"Human jaws are actually becoming smaller, the brain cage is increasing in size, teeth are losing cusps, and their roots are shrinking." By doing many skull tracings, Dr. Marshall formulated a projected scale of possible human skull changes to obtain an idea of how future man may look.

The models he created to exact dimensions for the Evolution of the Skull display at the anatomical museum suggest that future man will have a bald pate, squeezed, prominent features, and small jaws.

"Of course," Marshall admits, "we're controlling our environment now, and man wasn't able to do that in the past. Such things as genetic engineering could very well change these projections."

TV CENSUS If you lived in the world inside your television set, you'd find more Help Wanted ads for doctors and lawyers than for file clerks. Florid psychosis and amnesia would be more common than the flu, but accidents would almost never hurt.

"The universe is not to be narrowed down to the limits of our understanding . . . but our understanding must be stretched and enlarged to take in the image of the universe as it is discovered."

—Sir Francis Bacon

These are the findings of two psychologists: Alberta Siegel, of Stanford, who conducted a census of TV-land as part of the 1981 update of the 1972 surgeon general's report on TV and behavior, and Washington, D.C., psychologist Lorraine Bouthilet, who conducted a health survey of our cathode-ray-tube alternate universe.

Men outnumber women three to one on prime-time programs, Siegel says, and nubile females—mostly in their teens, twenties, or thirties—usually date men ten years older than themselves.

Harmony and exaggerated courtesy are the norm between spouses and between parents and children—which is odd since TV dads are so "inept, bumbling, absent, or incompetent." More moms are going into the business world now, and divorce is treated as an acceptable alternative.

Children and the elderly—TV's most faithful viewers—compose a minute fraction of the television population. In stark contrast to real-world demographics, old men are much more common than old women, and most older people are helpless and pitiful. Since children and the elderly use TV "as a window on a world otherwise inaccessible," Siegel wonders about the version of reality they receive.

While 7 percent of prime-time characters suffer grievous accidents, only a handful ever need to be hospitalized, Bouthilet reports. Yet they rarely buckle up for safety! But mental illness is epidemic: TV psychotics don't just pick daisies, either; most are violent or victimized.

In contrast to the physical robustness of prime-time protagonists, half of all day-time-serial characters are sick. Fortunately, TV doctors are almost uniformly ethical, fair, kind, young, successful, smart, and sociable. Also, unlike real physicians, who rarely make house calls, video M.D.'s do 60 percent of their doctoring outside the office.

Although TV folk snack about nine times an hour, extra adipose tissue never forms, nor does incessant drinking make most people rowdy, unstable, or uncool.

If your plants
start clicking at you, it
may simply be
a sign that they're thirsty.

NOISY VEGETABLES An Australian scientist has revealed that plants whisper, "Water me."

As St. Francis preached to birds, John Milburn, of the University of New England, at Armidale, New South Wales, has picked up the language of plants, according to the *New South Wales Business Review*. Eavesdropping on the castor bean plant—by means of a tiny microphone implanted in its stem—Milburn has heard thirsty plants *click.*

The clicking noises reportedly come from the vibration of the plant's minute water pipelines. His method, Milburn says, could tell farmers which seeds to plant in arid soil and identify which new breeds of plants are drought-resistant.

Other plant fanciers are underwhelmed by the breakthrough, however. "It's certainly plausible that the columns of water might snap after a while," says University of Chicago biologist Edward Garber. "But so what?"

What of Milford's claim that his plant-listening method could identify drought-defying plants? "He's out of his gourd," says Garber. "Say you came up with five hundred ways to tell whether someone is dead. That wouldn't bring him back to life."

MAGNETIC PEOPLE It's common knowledge that birds, bees, and even lowly bacteria are guided by a magnetic sense. But some scientists are claiming (to the annoyance of their colleagues) that people, too, can navigate with the help of magnetic perceptions.

The debate whether animal magnetism is present in humans began when University of Manchester zoologist R. Robin Baker blindfolded some of his students, packed them into buses, and set them on a twisting ride through the English countryside. Baker stopped the buses at various points and asked the students to guess where their starting point was and to indicate by pointing in a certain direction. Surprisingly, his subjects exhibited strong homing abilities.

The explanation, according to Baker, was their magnetic sixth sense. A few months later, however, several American researchers tried to reproduce Baker's results but failed.

Baker claims that his latest work "leaves no doubt" about the presence of a magnetic sense in human beings. He says that the American attempts were not really failures. Instead, he explains, the experiments may have been stymied by less than perfect conditions: confusing magnetic storms, the wrong time of day, and low magnetic gradients inside the buses.

To minimize these effects, Baker and his collaborator Janice Mather conducted tests in a specially constructed wooden hut. Blindfolded and earmuffed, students were seated on chairs in the middle of a dark room, turned around, and then asked to indicate their direction. After obtaining accurate answers from nearly 150 people, Baker is confident that he has proved his hypothesis.

Despite this latest research, Baker has yet to convince his American detractors. One has already called some of the data "a little bit funny."

THE BEDFORD PYRAMID Given enough money (and there seems to be enough of that) and about a years' more time, there will be a second Great Pyramid of Cheops, built this time not in Egypt but in Indiana.

The pyramid, to be made of Indiana limestone, will be erected on the 20-acre site of the Indiana Limestone Tourist and Demonstration Center, in Bedford. The cost: $700,000, which came to Bedford as a grant from the Economic Development Administration (somehow it escaped the Reagan budget scalpel).

The Indiana Limestone Tourist and Demonstration Center is a project designed to attract tourists to Indiana's quarry region. The site overlooks the two largest quarries in the country and is planned to be a combination museum and training center for stonecutters.

The quandary about what kind of building should be constructed on the

Egypt is known for its pyramids. Now the town of Bedford, Indiana, is mounting a challenge.

site was finally solved by Merle Edington, president of the Bedford Chamber of Commerce, who happened to be looking through a book, *Ancient Egypt: Discovering Its Splendors*, and saw the pyramids. Because of this inspiration, a replica, 96 feet, 8 inches tall, of the Great Pyramid of Cheops is being built in Bedford.

It is not an exact duplicate. The original is five times higher. And it is no indication that Bedford places any special stock in anything like pyramid power. "There's no mysticism involved, none of that malarkey," Edington stresses.

The site will also feature an 800-foot-long replica of the Great Wall of China, with towers, also made of limestone.

WHISTLING EARS One day at the Central Institute for the Deaf, in St. Louis, psychologist Patrick Zurek stuck a microphone into his right ear canal to investigate a rather routine aural phenomenon. To his surprise he heard a distinct, high-pitched, teakettlelike *whistling.*

Fascinated, Zurek tested 32 volunteers. Amazingly, half of them also broadcast sound from one or both ears. Most of the ear music was audible only by microphone, though a few subjects' ears carried on so loudly as to be heard by the naked ear. Biochemical events in the organ of Corti of the cochlea, Zurek says, are responsible.

A twenty-two-year-old Dutch woman was celebrated in a medical journal for ears so noisy that her sister complained of the din during their piano duets.

If Zurek's subjects are representative of the general populace, apparently half of us have noisy ears. Zurek theorizes that the phenomenon represents very subtle ear damage caused by pervasive noise pollution. A goodly number of chinchillas also betray the whistling-ear syndrome, Zurek reports, but chinchillas raised in quieter environments do not. If this data also apply to humans, it appears that silencing our environs will also silence our ears.

The Social Security pensioner of the future after cashing one of his monthly checks (he'll need it).

FOSSIL MUSIC Sure, they had fire, hunting and gathering, and puberty rites, but what did our Stone Age ancestors do for fun?

One answer comes from the Soviet Union, where a group of Kiev musicians has cut a record of caveman bebop, played on 20,000-year-old instruments.

It all started when six bone instruments—fashioned from the shoulder blades, hips, jaws, tusks, shanks, and skull of a mammoth—were found at a dig near Kiev. Criminologists, forensic doctors, and musicologists from Leningrad's Heritage Museum identified the artifacts as primitive drums, cymbals, and other percussion instruments.

The ancient instruments have not lost their tonal qualities; each has a characteristic sound. What did the Stone Age Top Forty sound like? The Kiev Academy enlisted musicians to find out, and the result is a long-playing disc on the Melodiya label, the Soviet Union's government-owned record company.

FUTURE MILLIONAIRES Children born in this decade who live a normal life span will probably all become millionaires if U.S. Social Security Administration (SSA) projections are correct.

Basing its forecast on an inflation rate of only 4 percent a year, the SSA said the average wage earner 70 years from now will collect $761,332 annually. Those on Social Security alone will receive an average benefit of $300,635, assuming the program still exists.

If you doubt those figures, you may wish to compute your own at a more accurate inflation rate. In 1980 the inflation rate ranged between 12 and 18 percent, and in 1981 it never dipped below 8 percent. The average wage today is about $12,000, and the average Social Security benefit is $5,682.

COINCIDENCE Improbable sounding coincidences are often put forth as evidence of thought transference, or ESP. The Nobel laureate physicist Luis

A man reads a newspaper, suddenly remembers a long-forgotten friend, and then finds out the person has just died . . .

W. Alvarez once was startled by a coincidence that happened to him.

Reading a newspaper one day, he came across a phrase that triggered associations and led to his thinking of a person from his college days, "very probably for the first time in thirty years." Five minutes later, in the same newspaper, he came across an obituary notice reporting the death of that person. Some persons would be quick to assume that clairvoyance or precognition was responsible.

Not Alvarez. He realized that the connection was a coincidence, and, like the good scientist he is, he proceeded to calculate the chances of its happening. He found that the probability of a coincidental recollection of a known person in a five-minute period just before learning of that person's death is about 3 parts in 100,000 per year. Multiplying by the 100 million adults in the United States, an incredible 3,000 such experiences of the sort should occur every year, or about 10 per day. Seemingly improbable coincidences, in other words, are more likely to occur than most of us would readily assume.

Alvarez concluded in a letter to the journal *Science* entitled "A Pseudo Experience in Parapsychology," "With such a large sample to draw from, it is not surprising that some exceedingly astonishing coincidences are reported in the parapsychological literature as proof of extrasensory perception in one form or another."

ELECTRIC FLYING CARPETS As part of his otherwise serious research into superconducting materials, Stanford University physicist William Little has suggested a whimsical use for superconductors: the flying carpet.

Speaking at a Quantum Theory Conference at the University of Florida, he proposed that all kinds of fantastic things would be possible with a superconducting material that operates at room temperature (present superconductors will work only at extremely low temperatures—hundreds of degrees below 0°F).

Surprise sight:
For the blind, the unseen
world can be
breathtakingly beautiful.

Using a sort of magnetic levitation, you could float cars over superconductor highways, transmit electricity thousands of miles with no loss, and, if you had a mind to, weave flying carpets, even flying suits, from the material.

First you would weave a large wall-to-wall carpet in a magnetic field, to trap some of the field in the carpet. You'd lay that down on the floor of the room and then weave a series of smaller rugs in another magnetic field. "You could use an Oriental design," Little says, perhaps alluding to Ali Baba. When these small rugs are laid over the larger one, they would float.

Each one, Little estimates, would hover about a yard above the floor and could float around easily, carrying a 200-pound person.

It might even be possible to use the same method to weave superconductor clothes. "Then you could fly around the same room," Little says.

Little often uses the flying-carpet example in his talks and lectures, and this usually interests everyone. But there are some exceptions. "When word of this first came out," he recalls, "I got letters from practitioners of Transcendental Meditation, who wrote something like 'What's the big deal? We've been doing this for years.' "

SURPRISE SIGHT A routine cataract operation on a sixty-two-year-old California woman blind from birth has shown that the brain can often fill in what the eyes cannot see. After a cataract had been removed from her left eye, Anna Mae Pennica astounded some vision experts by announcing that the world of sight was "pretty much as I expected."

Shortly after the operation, she was able to read print and recognize certain colors, such as green. The print she knew from having been drilled in tracing the shape of letters as a young girl. As for green, she explained that she often dreamed in colors.

Dr. Thomas Pettit, the UCLA surgeon who performed the operation, said he knew of no other case like Mrs. Pennica's where someone born blind was able to see so easily without experience.

A similar case did occur in the 1950s, according to Georgetown University psychobiologist Dr. Richard Restak. An English shoe repairman, known only as S.B., who had been blind since the age of ten months, had his sight restored by an operation when he was fifty-two years old.

He quickly began reading and doing other things, such as telling time, but, according to British psychologist Richard Gregory, who studied S.B., the man, unlike Mrs. Pennica, found the world "a drab place." Formerly an exuberant person, the man became irrevocably depressed by his disappointment that the world was not as he had imagined it.

Since there is a running debate among theorists about whether a person can learn to see with little or no vision during the developmental early years, vision experts have been recommending that Mrs. Pennica's recovery be given special attention.

PLASTIC BEES When Mama bees of the Colletes genus lay their eggs, they take care to protect the offspring from environmental hazards. So, with a secretion from their bodies, they form a waterproof, wear-resistant brood sac to hold the eggs and equip it with a stash of pollen and nectar for the bees-to-be.

Researchers analyzing the sac were surprised: The material the mother manufactures is a natural type of polyester.

The female releases the plastic through a gland on her abdomen. Then she molds the thimble-sized sac (resembling a resealable sandwich bag) with her tongue and legs. Wrapping her babies in the petrochemical-free plastic, she proceeds to bury the brood in an underground nest.

Suzanne Batra, an entomologist with the U.S. Department of Agriculture, discovered the polyester sac while studying the wild bees. She notes that these are the first species found to make plastic. Other types of Colletes produce a brood sac of silk, similar in appearance to the plastic pouch. However, neither of these bees can make honey.

> *"Stand firm in your refusal to*
> *remain conscious during algebra.*
> *In real life, I assure you, there is*
> *no such thing as algebra."*
> — Fran Lebowitz

According to an evolution theory that she developed, Batra contends that the plastic- and silk-secreting bees are less "evolved" than the honey producers.

Will the lowly polyester-producing critters someday be churning out men's leisure suits? Batra says that further research is needed before it's known whether the plastic has commercial potential.

TASTY GENES The reason people differ in their taste for soft drinks may not be all in their head. It might be in their genes. It appears, according to Linda Bartoshuk, of Yale University's Pierce Foundation, that people's tongues may be as different, physiologically, from one another as their brains are.

"We've been misunderstanding certain facts about human behavior in taste for a long time," says Bartoshuk. "You tend to assume that everybody can taste the same thing and that some people can take it and some people can't. Well, that's simply not true. The experience people have is quite different."

The ability to taste certain bitter compounds, the researcher has found, is determined by our genes. The genes design the receptors that will appear on the surface of the tongue. She has discovered that there are actually separate receptors on this muscular organ that respond to quite different groups of molecules. Phenylthiocarbamide, or PTC, is bitter to some but tasteless to others. Yet people who are either sensitive or insensitive to PTC can taste quinine, another bitter substance. The conclusion: Because PTC nontasters are sensitive to the bitterness of quinine, there may be at least two distinct bitterness receptors on the tongue.

What does this mean to the soft drink and food industry? Bartoshuk believes that the study may allow researchers to develop combination sweeteners that do not leave a bitter taste in the mouth. But she has her doubts about whether it will be followed through. "Companies that develop sweeteners want to push only one molecule because of the expense of testing and

Unlike humans,
plants cannot cry when suffering
from stress, yet
they show ill effects all the same.

because they don't want to share the profits with other companies," concludes the researcher at Yale.

PLANT STRESS Scientists have known for years that plants, as well as animals, suffer from the ravages of stress. These stresses may come from too little light, lack of water, or a paucity of nutrients. Now researchers at the University of California, Davis, have developed a cure for at least some of the plants' stressful situations.

Raymond Valentine and his associates have isolated a gene from the DNA molecule that may someday help farm crops withstand such stresses as drought and high salt content in soils. Years of irrigation, for instance, can destroy the essential nutrients in the dirt and build up potentially harmful substances in their stead. The gene would improve and maintain crop yields by making plants more resistant to these factors.

Although Valentine has not been able to apply this procedure to higher plants (the operation is too complicated), he has been successful with bacteria, the lowest organisms in the plant kingdom. The team transferred the gene that controls the overproduction of a certain building block of protein in one bacterium and placed it into another. Overproduction of the amino acid, called proline, enables bacteria to withstand an unusually high salt content in their environment.

The host bacteria with the newly introduced proline were able to function perfectly well in the salty soil. The results prove, says Valentine, that it is possible to genetically engineer stress tolerance in microbes.

The research has far-reaching implications. The once-fertile lands of India were destroyed in part because of the accumulation of salt in the area. Today, says Valentine, a similar threat hangs over the heads of farmers of the Great Central Valley of California, one of the most productive regions on Earth. Such genetic study could give new life to these and other food crop areas around the world.

Victim of Pompeii:
Studies at Mount St. Helens show
that most volcano
victims die from suffocation.

VOLCANO DEATHS The explosion sent tons of stones and hot cinders down upon the fleeing Romans. Within a few hours, Pompeii was buried under 20 feet of ash. Yet for some 1,900 years the question of how the victims of the Mount Vesuvius eruption died had never been fully answered. But results of a study of a more recent volcanic tragedy may shed some light on how volcanoes kill.

Several months ago, John Eisele, at the King County Medical Examiner's Office in Washington state, and others began to piece together information about the victims of the May 18, 1980, eruption of Mount St. Helens. The researchers examined 25 of the victims. A few, they discovered, had died from head injuries caused by falling rocks and trees. A few had burned to death. But the overall majority had died of suffocation.

The culprit was volcanic ash. The bodies were soaked in it. Even routine incisions into the skin dulled scalpel blades. But the fine, yet gritty powder had its killing effects within the victim's throat. The ash, sucked in by the panicked campers, combined with mucus and formed a thick plug in their air tubes. Death occurred within minutes.

Others, caught when the superheated water rushed down the volcano's slope, died of burns. Their bodies, unlike those of fire victims, were not charred but appeared dried and baked. The skin was mummified; the exposed muscles were dehydrated and frayed.

The victims were not frozen in place, as early accounts had described. They did have time to escape before the ash came. Several people with burns were found miles from the scene. And film, taken from a camera of one of the dead, showed plumes of ash looming overhead. The pictures had been shot from the rear window of a moving truck.

Eisele believes that some deaths might be prevented if people in the region of a volcano use gas masks or stand under adequate shelter when an eruption occurs. But, he adds, nothing can protect those unfortunate enough to find themselves at the base of an exploding volcano.

If Limburger scientists wanted an expert opinion, they should have brought in this consultant as head of research.

SNIFFING OUT LIMBURGER'S SCENT One of the lingering questions facing food scientists for many years has been what makes Limburger cheese smell so awful. They knew that growth of yeast and micrococci *Brevibacterium linens* created Limburger's flavor, but the chemical components of that room-clearing redolence remained a mystery.

A scientific team at General Foods Technical Center, in White Plains, New York, may have unlocked the secrets of smelly cheeses. For better or worse, team leader Thomas H. Parliament believes he can re-create Limburger aroma in an artificial cheese.

Parliament knew that the aromas of many cheeses derive from particular chemicals. Blue cheese can thank a family of methyl ketones for its tangy odor. Methyl disulfide, which gives off an oniony odor on dilution, lies behind cheddar cheese's smell. Limburger smell, it turned out, is an incredible mélange of smell chemicals, far more complex than normal cheese.

Parliament's team found chemicals with floral aromas, the smell of tar, sulfurous-smelling compounds, potato- and cauliflowerlike odors, and chemicals that smell sour, rancid, and sweaty. In short, it appears that science has unlocked the secret of Limburger cheese's smell: It is composed of equal parts of the aromas of a dirty kitchen and an old gymnasium.

CHANGING ACCENTS In the not too distant future, Americans are going to sound more alike than they do today, an Ohio State University linguist believes.

Melanie Lusk has conducted language research among multigenerational families to observe accent changes from grandparents to parents to children. Sophisticated sound-recording and sound-analysis equipment and computerized data analysis make it possible to measure very fine nuances and changes, Lusk says.

The results? Linguists are observing "a living, changing language," Lusk says. "We know people don't go to bed one night and say, 'tomorrow I'm

If an Ohio State linguist is right, these New York kids may soon be talking with a Southern drawl.

going to change my speech.' Now we can watch it happen."

Specifically, Lusk notes, her research shows that features of northern accents are moving south, features of southern accents are moving north, both are moving west, and they seem to meet in the middle of Oklahoma (near the natural linguistic boundary of Route I-70).

Although a few sounds remain remarkably stable across three generations, many others are changing noticeably among teen-agers. In particular, children seem to be adopting the "characteristic New York 'a' sound," in such words as *bad*, saying them as "baehd." They are also adding an additional "a" sound to the "o" in such words as *road* and *home,* saying them as "raoad" and "haome," again a northern shift.

Yet another common change Lusk has found is the merger of two distinct sounds in many words. "Whole word classes are collapsing into one sound: *pod* and *pawed; dawn* and *Don; caught* and *cot.*"

How do these changes occur? Individual contact is one way, Lusk says. "Someone knows someone else twenty-five miles away, who knows someone else twenty-five miles away, and so on."

But she has another answer, too. "Teen-age girls are a remarkable language lab. They pick up fads in language just as they do fads in clothes and records. They're just a few years away from being mothers. This helps us understand how language changes."

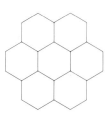

INVENTIONS
AND INNOVATIONS

CHAPTER 8

After one or two puffs,
the smoker "spits, spits, spits"
and the mouth
feels like a garbage dump.

FOUL MOUTHWASH A new mouthwash makes smokers as objectionable to themselves as they often are to nonsmokers. According to the inventor, Dr. William Najjar, a Montreal physician, the mouthwash has a fresh, minty taste that makes the mouth feel extraordinarily clean—until the smoker lights up. Then the mouth tastes "as if it has chewed on fifty cigars, and the smoker is instantly in hell."

Dr. Najjar reports an 85 to 90 percent cure rate for smokers who use the mouthwash in a ten-day regimen. A smoker merely rinses twice daily with one teaspoon of the mouthwash in an ounce of water; the devastating effects last up to 18 hours. Dousing the mouth with silver nitrate and glycerine, the mouthwash produces a bracing sensation, drying and tightening the gums and activating the salivary glands. When the smoker takes a drag, the tar and nicotine absorbed by the gums react with the mouthwash and saliva, forming a bilious yellow liquid. After one or two puffs, the smoker "spits, spits, spits," and the mouth feels like a garbage dump.

Trained in neuropsychiatry, Dr. Najjar got sick of treating patients with smoking ailments. Finding rational arguments useless, he resolved to cure them with a "shock therapy." Without warning, Najjar gave the mouthwash to a few patients and let them go their merry way. It made their breath unbearable. Knowing he'd hit on something, the doctor spent the next seven years testing the rinse on some 40,000 subjects.

Dr. Najjar has begun to sell the mouthwash in Quebec and Ontario provinces under the name of Tabanil (a French portmanteau word for "tobacconil"). He is fighting for Food and Drug Administration (FDA) approval so he can sell it in the United States. When and if it gets here, potheads should be alerted as well: The mouthwash has the same effect on people who smoke marijuana and hashish.

ROCK AND ROLL HOT PANTS The latest discotechnological breakthrough for these unsettled times is an item called Rock and Roll Hot Pants.

Hot pants: An incredible tingle from rock music, but the ultimate thrill is the *1812 Overture*.

By wiring your shorts or panties to a stereo speaker by means of a 15-foot cord, which relays the music to a two-inch disc on your waistband, "you get an incredible tingle all over your body," claims inventor David Lloyd.

Lloyd got the idea for the stereo hot pants, which come in both men's and women's versions, from a Newton-like, serendipitous incident. Sitting under a previous invention, a stereo "flying saucer that hangs from the ceiling," Lloyd jumped when it fell in his lap. "Then, de doop, de doop, doop, doop, I began to smile," Lloyd says. "It felt good. I said, 'Wait a second, I think I've just invented something.' We worked on developing a really pulsating speaker, and now it's going well. We thought at first it was almost a joke, but the response has been fabulous."

For those who want the ultimate thrill, Lloyd recommends not rock and roll but classical music. "Try the *1812 Overture*," he says. "The cannons are something else."

CONSPICUOUS DRUNKS Separating the Perrier sippers from the truly wasted is becoming a simple matter in Los Angeles, where drunk drivers can be spotted miles away.

A group of second-conviction drunk drivers is driving with a built-in sobriety test. Whenever the ignition is switched on, the driver must pass a steering-competence test. If he fails and drives anyway, the car's emergency lights flash and the horn honks repeatedly.

The Drunk Driver Warning System (DDWS) also has a cassette recorder that monitors driver use and records the times when a driver fails the test or drives with the alarms activated. The system costs $500 to $600.

The purpose of the DDWS is not to rehabilitate the drunk, says Thomas G. Ryan, chief of the Alcohol Impairment Group Problem Behavior Research Division of the National Highway Traffic Safety Administration. "It's to protect him from himself and to protect the public from him, while still allowing him to drive, especially if he needs the car for his job," Ryan explains.

When a driver fails the steering test, he must wait ten minutes before he can repeat it.

SPEECH FOR THE DEAF Deaf children as young as three are learning to speak by using their sight and a new computer-aided system called dynamic orometrics.

Developed by a team led by Samuel G. Fletcher, of the University of Alabama in Birmingham, dynamic orometrics is a method of measuring the movements of the tongue, lips, and jaw. It works by translating these movements onto a television screen and showing the deaf pupil how the movements he or she is making for a specific word compare with those of a hearing person.

A custom-fitted pseudo-palate, a thin, vacuum-molded plastic plate, is placed in the mouth of a deaf child and of a hearing adult. Whenever the tongue moves or touches the pseudo-palate, it activates sensors embedded in the device and the position of the tongue appears on a television screen superimposed on a stick profile of the head.

This technique also allows the deaf to visualize pitch, volume, and the position of the lips and chin. A square drawn in the throat area of the stick profile rises or falls with changes in pitch and widens or narrows to indicate changes in volume. Light sources on the lips and chin are used to show movements of the oral structures outside of the mouth. This way, the hearing-impaired person can see that the lips are placed close together to say "see" and far apart to say "of."

"With this kind of access to the previously hidden actions of speech, persons as young and handicapped as a three-year-old child with profound hearing impairment can see and learn the intricate actions of articulate speech," says Fletcher.

Once the deaf know the movements of speech and the sensations of the movements, they no longer need to see what they are doing on the screen.

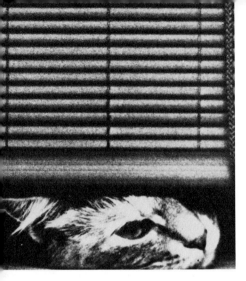

Ginger-colored felines
have two major enemies: black cats
and nuclear warheads.
A new device controls both.

"They don't have to carry a computer around on their shoulders," says Fletcher.

Further studies in dynamic orometrics are being funded by the National Science Foundation.

CAT-FLAP BOMB Cat fanciers and proponents of nuclear disarmament now have a way to satisfy both of their causes at the same time.

In the files of the English Patent Office is patent number 1 426 698, or "Photon push-pull radiation detector for use in chromatically selective cat-flap control and 1,000-megatonne earth-orbital, peacekeeping bomb," as it is officially known. According to the English publication *New Scientist,* the idea was the brainchild of a now deceased Patent Office employee who was looking for a way to feed his aging cat and, by a flash of insight, also felt he stumbled onto the ultimate nuclear deterrent.

The idea came after the inventor kept seeing his old, ginger-colored cat "overtaken on the way to his cat food by the black cat from next door, who is much younger and more agile."

His solution was a special, light-sensitive switch that opens the cat flap in the door for a ginger cat but not for a black one. As a cat approaches the flap its weight triggers the switch, which flicks on two lights. A photoelectric cell measures how much light is bounced off the animal and will open the flap only if a great deal of light is reflected—as from a light-colored cat.

In a related patent for an ARNDS (Automatic Response Nuclear Deterrent System), the same inventor says the cat-flap switch could be hooked up to a 1,000-megaton bomb put in orbit around the earth. When it senses the light given off by a nuclear-missile launch from any country, the orbiting gadget would, instead of opening a cat flap, drop the bomb on the country.

In defense of his 1,000-megaton cat flap, the inventor said: "If all your nuclear energy was used for peaceful purposes, instead of a large part of it being stored for blowing each other to bits with H-bombs and the like, you

> *"Nothing is impossible for the
> man who doesn't have to do it
> himself."*
> —*Weller's Law*

could save a hell of a lot of money, which would help to stop world inflation and," he added, "might even bring down the cost of tinned cat food."

RAINY-DAY GENES If you preserve a sample of your DNA for 50 years, future scientists may use it to save or extend your life. That's the optimistic conviction of the DNA Security Company, in Fairfax, Virginia, now selling a do-it-yourself DNA storage kit. The kit includes a small lance to produce a drop of blood from a fingertip, a plastic storage container, and a documentation form on archival paper.

To use, simply prick your finger with the lance and drain a few drops of blood (whose cells contain your genetic code) into the container. Then send the container, along with the form, to DNA Security offices in Fairfax. For a fee of $1 a year, the company will store the DNA in a fireproof bank vault in northern Virginia, where earthquakes are rare.

When genetic engineering is applied to human cells, Earl May, the president of DNA Security, believes that scientists will be able to repair body organs, reverse the effects of disease, and eventually retard the aging process. "However, since the quality of genetic information deteriorates with age, storing youthful, high-quality cells may mean the difference between life and death."

DNA Security developed the kit after consulting with scientists at the University of Maryland, the National Institutes of Health, and the National Archives. University of Maryland microbiologist Lore McNicol, for example, agrees that "the procedure recommended by DNA Security would probably yield usable DNA for several hundred years."

Other scientists, however, contend that the idea of using DNA to repair your body organs might be a little farfetched.

TALKING CHECKOUT If the checkout counter starts talking to you the next time you visit the supermarket, you're not hallucinating. It's just the latest use

The endless chattering
of talking checkout laser scanners
may make you long for
the soothing banality of Muzak.

of electronic technology to imitate human speech.

Electronic circuits that can generate signals that, when fed through a speaker, become intelligible words have found increasing use during the past few years. One example is the Speak and Spell electronic toy, made by Texas Instruments, Inc.

Another maker of speech-synthesis circuits, the National Semiconductor Corporation, of Santa Clara, California, also makes laser scanners that automatically read the striped codes on food packages at supermarket checkout counters. So it was probably inevitable that the two technologies would get together.

Tests of talking scanners have already begun in a San Jose, California, supermarket, and the first permanent installations were due to begin as this book went to press.

The speech circuit is intended to tell customers the price of each item they buy. With a 274-word vocabulary, the circuit can also be programmed to say such things as "Thank you for shopping with us today." The speech capability "will bring back an old friend to the counter," proclaims a National Semiconductor press release.

However, the endless chattering of a bank of talking scanners might make customers long for what one observer called "the soothing banality of Muzak." So might the clerks who would be left with the traditionally mechanical job of moving things while the machine did the talking.

CAR FINDER For the absent-minded driver who can't recall where he's left his car, there's hope at last.

A Torrance, California, firm is marketing an antitheft device, called Pulsafe, which—at the press of a button—will make your car beep its horn and flash its headlights.

Perfect for those mammoth parking lots at the airport after a weekend trip, or at a sports complex after a beered-up afternoon, Pulsafe uses a radio

195

transmitter and operates on the same principle as an automatic garage-door opener. Just flip a switch, and your car springs into action. The optimal signal radius is 100 feet.

"The device is also useful in unsavory neighborhoods for scaring off young people who have selected your vehicle as a backstop for stickball," says Marvin Lazansky, president of TMX, Inc., the distributor. When the car goes crazy with no one in it, the kids just scatter. The suggested retail price is $495, not including installation.

NEW SUITS FOR G-MEN As jet aircraft streak to ever greater speeds, methods are needed to protect their pilots against body-jarring sharp turns and dives.

A new rapid-acting gravity protection system has been developed for the navy to be used by today's fighter pilots, who can receive over eight G's, or eight times the pull of Earth's gravity, during acceleration and rapid maneuvering.

Such abrupt forces cause blood to pool in the legs and diminish in the head, resulting in discomfort, pain, or possibly leading to a pilot's blacking out and losing control of the aircraft.

G-protection suits currently employed rely on inflatable bladders to exert pressure on the abdomen and lower-body parts of a pilot, hence controlling blood flow. Although the suits are widely used, they are sluggish in reacting to G-forces and use excessive inflation pressure, creating unnecessary discomfort and an impairment in pilot performance.

According to co-inventor Richard Crosbie: "This new system is intended to match instantaneously the pressure in a pilot's suit with the acceleration value of the jet plane. Also, the electronic sensing system automatically adjusts the tilt angle of the pilot's seat, enhancing the ability to tolerate greater G-levels for longer durations."

Manufacture and adaption of the system to high-performance aircraft are

To avoid his chief nemesis—the dog—as well as other dangers, a postman has invented a movable mailbox.

both expected to be relatively simple. The new system could also be applied to future tourist-class space shuttle orbiters, states Crosbie.

MOVABLE MAILBOX When mailman Raymond Vanis crossed the sprawling lawns of David City, Nebraska, he invariably slipped on ice, tripped over toys, or provoked dogs to attack. After more than 15 years of jangled nerves and mounting injuries, he decided to tackle the problem. The result? The Curb-to-Door Mail Retriever, a robotlike arm that will help postal carriers transport mail across a yard.

According to Vanis, his new device consists of a folding mechanical arm with a portable mailbox at the end. The arm stands in the middle of a yard, Vanis explains, and is hooked to a pulley system that extends from the curb to the front door of the house. By tugging the pulley cord, the postman and the receiver of the mail can move the arm and the mailbox back and forth without braving the hazards of the lawn.

Vanis expects that if his system is implemented, one mail carrier could do more work than is now done by two, thus reducing manpower costs. Moreover, the carriers who remain would require far less time off to nurse their bruises. As an additional incentive to homeowners, Vanis has designed the device to double as a support for a flag, yard light, or basketball hoop.

TALKING THERMOMETER If you have trouble reading the red line on thermometers, Electromedics, Inc., a Colorado company, has just the thing for you: a talking instrument that actually tells you your body temperature.

Don Nicholson, of Electromedics, says, "We developed the instrument for the blind, but it is being used primarily in biofeedback work. Once the probe is attached to an extremity, a finger or a toe, the thermometer gives continuous readouts in a male voice." Oral and rectal probes are also available. The talking instrument, which Nicholson said lisps the numeral 3, saying "thwee," retails for about $270.

The new computer
program can help this prospective
homeowner determine
whether renting makes more sense.

A less expensive, nontalking model that gives a continuous digital readout of temperature is more popular, at about $100, Nicholson says.

"Both are useful in biofeedback because they tell the user how he's doing. As you relax, for example, the temperature in the extremities goes up."

BUY OR RENT? Those who are deciding whether to buy a home or to rent one can get a little scientific advice on the matter from Michael Johnson's computer program.

Johnson, as assistant professor of consumer economics at the New York State College of Human Ecology, has developed a program called Equivalent Rent Analysis, which figures in all the financial variables. For $5, Johnson will tell a renter thinking about taking out a mortgage whether it makes financial sense.

Once sent the fee, Johnson mails out a detailed questionnaire, asking for present monthly rent, net income, tax situation, local mortgage rates, closing fee, utility and maintenance costs, and expected appreciation of property.

The program analyzes the running costs of homeowning—mortgage payments, property taxes, upkeep, for example—how much of a profit or gain there would be from selling the house after a certain number of years of ownership, and finally how the gain, if there is one, breaks down on a monthly basis. And that is the critical rent-or-buy factor, Johnson says.

Some real estate programs already exist, but Johnson, who's spent one and a half years developing his, says many tend to be biased and none look at renting as a real alternative. Nor do any work with three separate inflation scenarios of bad, average, and best, as his does.

For more information write: Equivalent Rent Analysis, c/o Professor Michael S. Johnson, 108 MVR Hall, Cornell University, Ithaca, N.Y. 14853.

SPYING ON THE WRIGHT BROTHERS Had a turn-of-the-century spy been just a tad more prescient, the Wright brothers' good old American

"God has infinite time to give us;
but how did He give it?
In one immense tract of lazy
milleniums? No. He cut it up
into a neat succession of new
mornings."
　　　　　—Ralph Waldo Emerson

know-how would have gone British, says a University of California at Santa Barbara historian.

From unpublished documents in British and American archives, Alfred M. Gollim—the author of a recent book on British air power—has pieced together the remarkable tale of a Briton who spied on the Wrights for a year but missed a crucial rendezvous.

One Patrick Y. Alexander, a member of the Aeronautic Society of Great Britain, became friendly with Wilbur and Orville Wright in the pre-Kitty Hawk days, when they were testing gliders in Dayton, Ohio. Though he was invited to Kitty Hawk, North Carolina, in December 1903, the would-be spy "got mixed up" and was in the wrong place to witness man's first flight, according to Gollim.

As it was, five years elapsed before the U.S. government stopped scoffing at the invention and bought a flying machine. Britain got one in 1908, too, and its designer owed much to the prior American invention.

Had Alexander been on the spot to see how the flying machine worked in 1903, Gollim says, Britain would probably have gained several years' lead over all other countries in aviation. And, for one thing, the Royal Air Force might have settled World War I much more quickly.

DOUBLE VISION Using a device small enough to fit on a pair of eyeglasses, the deaf may be able to read lips without looking up and helicopter pilots may be able to read their instrument panels without looking down.

The device is called an Eyeglass Heads-Up Display. It works by taking light from a little packet of optical fibers and bouncing it off a tiny mirror stuck in the center of an eyeglass lens and into the user's eye. Because the mirror is so small and so close to the eye, it is virtually invisible. And the image it reflects appears to be superimposed onto the user's field of vision.

The device was originally conceived by Hubert W. Upton, group engineer for electronics research at Bell Helicopter Textron, in Fort Worth, Texas, as

The radio transmitter of the future will beam a second signal that carries information along with the music.

what he calls a visual speech reader for people with hearing difficulties.

A computer takes a distinctive sound—the ess sound, for example—and projects it as a specific color to the eye. By memorizing what colors stand for which sounds, one can translate the colors one sees into words. Afflicted with a hearing problem himself, Upton now uses the device and says adapting to it is "kind of like learning a foreign language."

The U.S. Navy is also testing it for use with night flying goggles, which its helicopter pilots wear and which restrict a pilot's view of his instrument panel. By hooking the Heads-Up Display to an on-board computer, a pilot can have information from the control panel projected to his eye.

According to engineer James R. Goodman, of Textron, who helped design the display system, future applications might include special safety glasses that could let people operating dangerous machinery monitor their instruments without lifting their eyes from their work.

FUTURE RADIO Just when it seems as though television will be riding high on the wave of the future, the British Broadcasting Company (BBC) in London has come up with a revolutionary idea for the radio. The BBC has announced that it is experimenting with what it calls radio-data—information that will appear on a display on the front of your radio receiver when the set is turned on.

Ordinarily, a radio station transmits on one wavelength signal. That signal carries everything from music to mindless chatter from the station to the receiver in your home. But the new transmissions would carry a second minor signal along with the regular wavelength. While you are listening to your favorite music, information would be flashing across your receiver display just as it does when you're watching a show on television and an announcement appears at the bottom of the screen.

The additional signal wouldn't affect the regular signal but would activate the display to provide sports scores, channel identification, and possibly

*"I must confess that my
imagination . . . refuses to see any
sort of submarine doing
anything but suffocate its crew and
founder at sea."*
—H. G. Wells, 1902

some news headlines—all while you are listening to your favorite station. "The idea," says a spokesman at the BBC, "is to identify the station you're tuned to and, if possible, to include a very brief synopsis at the start of the program."

If the tests prove successful, listeners will be able to buy the special receivers at their local stores within a few years. These receivers will have the ability not only to display information but to tune automatically to a station playing a particular style of music. "On a sophisticated receiver," explains the BBC spokesman, "you could program the set to look only for a certain type of pop or country and western music. Each program that came along would have a unique code. If a program was transmitted for country and western music, the radio receiver would scan, lock onto that unique code, and play it."

So far, the only nations investigating this new radio idea are England, Sweden, and France. But if the idea catches on in the United States, you may be able to receive sports scores on your set within a few years.

HOLOCAUST HIVE Now you and your honey can be snug in your prefab fiberglass hive while above you the world blasts itself into nuclear smithereens. The reason for this happy situation is the Honeycomb Bombcell, a high-tech bomb shelter manufactured by Design for Defence, in England.

The 14' by 11' Bombcell folds flat for easy shipping, costs one-third less than a concrete bunker, and can be dropped into any appropriate-size hole with ease. With four bunks, shelves, cupboards, shower, and a toilet that shoots waste upward to the polluted surface, the cells will keep you comfy from a one-megaton bomb 2.5 miles away or a five-megaton burst 3.75 miles away.

Daylight filters through two water-filled, radiation-soaking tubes. Air freshness is maintained through a pedal-power pump that also keeps the occupants in good shape for the anticipated return to the surface. Beneath the

Now you can select
a vintage that features an
excellent bouquet and
non-fermenting microorganisms.

floor sits 420 gallons of water in hermetically sealed tanks.

You can even get a pop-up TV camera to give you a view of the carnage above from your subterranean hearth.

WINE FOR TEETOTALERS You're at a party and someone offers you a glass of wine. Unfortunately, your doctor has told you that you can't have any alcoholic beverages. What do you do? Well, you can decline the offer and ask for bottled water or soda. Or, when it comes out on the market in the very near future, you can ask for the nonalcoholic wine and join the rest of the drinkers.

Two researchers at the University of Tel Aviv, in Israel, have developed a natural, nonalcoholic drink that tastes like, looks like, and smells like the real thing.

Henry Margulis and Avraham Lifshitz designed the drink for those people who for medical, religious, or ethical reasons can't imbibe alcohol. "One may also use it for certain sick people, children, teen-agers, and supporters of the Anti-Alcoholism League and other similar organizations," adds a spokesman for Ramot, the university authority that is in charge of the commercialization of the researchers' product. "This drink may also be enjoyed by people who are not used to drinking wine but still desire to have a light drink with their dinner."

The process for making this beverage is different from that for traditional wine making. Wine results when grape sugar is broken down into carbon dioxide and alcohol. But Margulis and Lifshitz's drink is made with microorganisms that do not complete the fermentation process. Sugar is broken down, but not into alcohol and carbon dioxide. Afterward, the liquid is filtered, pasteurized, and bottled.

The "wine," which comes in dry and semi-sweet red, white, and rosé, is expected to cost a quarter of the price of regular wine. It should be out on the European market soon.

Even the most
futuristic skates share one
feature with
old-fashioned models—no brakes.

SUPERSKATES Back in the good old days, roller skates were simple. You strapped them onto your shoes, tightened them with a key, and rumbled off down the street. No longer. The skating craze has brought specialization. There are disco skates and racing skates and casual skates. And now we have Techno-skates. Freewheeler Leisure Products, in England, has produced a skate that tries to incorporate the best in technology and modern design.

The Mark II Roller Skate chassis is formed of heat-treated aluminum alloy shaped into a broad flat support for a jogging shoe. It has axles that can accept broad wheels for street skating or narrow wheels for better pinpoint dancing. And the torque on each wheel can be adjusted to suit the user's task and technique.

One thing Freewheeler couldn't engineer into their skate: brakes. If you tangle your toes, you'll still hit the deck.

CUSTOM TV ADVERTISING The informational power of computers may or may not someday create the sublime reality of an ultimate library. But it is already creating a spooky system of ultimate advertising.

In a half-dozen cities right now, consumers are being fed custom-tailored television commercials based upon a wealth of information about their lives and buying habits that has been collected and evaluated by computer.

The system begins at the checkout counter, where goods are scanned by those computer-laser cash registers. In test towns, such as Midland, Texas, and Evansville, Indiana, volunteer subjects have their every purchase piped to a central marketing computer. The computer evaluates their buying patterns and selects television commercials to be sent over their home cable-TV systems.

For example, a family that buys lots of cereal might receive ads for a new super-sugar-popping-three-grain granola, whereas another family, with a five-year-old car, would get auto ads instead.

> *"Science is built up with facts, as a house is with stones. But a collection of facts is no more a science than a heap of stones is a house."*
> —Jules Henri Poincaré

The experimental practice is growing ever more popular with marketers as computers become more ubiquitous and cable systems proliferate. "It's the answer to a maiden's prayer in terms of market technology," says Dudley Rich, vice-president for market research at Quaker Oats.

At the same time, it disquiets civil libertarians. "I find it shocking," says Norma Rollins, director of the privacy project at the American Civil Liberties Union in New York City. "Trying to beam commercials specifically [into individual homes]—I see that as a very large invasion of privacy.

"It gets tiresome sometimes to say it's almost 1984," she adds, "but it is."

REINVENTING THE OXCART One of the most vexing problems for technologists has been developing viable new systems for the Third World. For the poor, uneducated masses of these countries, fancy gadgetry doesn't make the grade. A successful system must strip a technological concept down to its most elemental application.

But even when that is done successfully, unexpected problems can ruin a good idea.

Take the case of S. Balaram, of the National Institute of Design, in Ahmedabad, India. He realized that the ancient creaking ox carts used by India's merchants and farmers were robbing the people of a vital energy source—a healthy, strong ox. The old carts mercilessly overworked the beasts, which carry 100 million tons of goods yearly, second only to the railroads.

So Balaram devised a new kind of cart that could increase productivity enormously by reducing strain on the pack animal. It had three wheels instead of four, with pneumatic tires in the back. He also suggested putting rubber pads from cut-up old tires on the bull's hooves, to give him better traction and cushioning on rough roads.

Sounds good? Sure, but in test practice the carts created instant shortages of rubber tires, as farmers cut up everything in sight for the oxen. And it turned out that the rubber tires gave more friction as well as more traction, so

205

India's ancient
oxcarts rob merchants and farmers
of a vital energy
source: a strong, healthy ox.

the ox, while he was more comfortable, actually worked harder to move his load. Balaram is back at the drawing board.

AQUA PURA MOBILE If Sam Leslie Leach weren't already a millionaire inventor he surely would never have gotten his idea off the ground. Leach proposed, in the late 1970s, that he could build a car that ran on water.

Impossible, snorted skeptical scientists. The car would take more energy to run than it could produce. It would sputter to a halt after a few feet. His notion of splitting water into hydrogen and oxygen, then burning the hydrogen, violated the laws of physics, they argued.

Be that as it may, Leach began getting patents on his system. He sold development rights to the Presley Companies and saw their stock rise from $3 to $20 a share.

Now, Leach is testing his "impossible" process in a 1980 Plymouth Horizon outfitted with an SLX (for Sam Leach Experimental) reformer in the trunk. This device breaks the water into hydrogen and oxygen and funnels the hydrogen to the engine.

"It's a beautiful running car," Leach reports, based on lab tests with the engine. He thinks, after road tests, that he will be able to demonstrate a water-fueled car that gets 12 to 15 miles per gallon.

And, if your car breaks down in the desert, you can always drink the fuel.

CELL CINEMA Blood platelets simmer like geyser pools, and molecules zoom down tubes faster than traffic on a freeway. This pulsating movement, taking place within the living cell, can now be observed with the help of a new video microscope.

Built by Dartmouth biologist Robert Allen, the new system has two parts: a television camera and a powerful optical microscope. According to Allen, the device "works on simple principles that somehow went unobserved for years." He explains: "When we set up the ordinary microscope for visual

*"The man with a new idea is a crank
until that idea succeeds."*
                        —Mark Twain

observation, we normally arrive at a trade-off between high contrast, with the aperture closed, and high resolution, with the aperture open." This trade-off results in an image that just isn't lighted enough to reveal rapid movement.

But, says Allen, "the TV camera offered a way around that obstacle because it sent all pictures to a TV screen in the form of an electric current. Allen found that by modifying that current *in transit,* before it reached the TV screen, he could raise both the contrast *and* the resolution to high levels. Allen was thus able to generate TV images that were bright and clear enough to reveal the intricate movements of a single cell.

Allen is currently using the new microscope to study microtubules—spaghetti-shaped fibers that support the cell the way bones support the human body. His recent video tapes reveal that microtubules, once thought to be immobile, actually snake through the cell, transporting chemicals. This is just one of the many discoveries Allen may make with his new microscope.

DO-IT-YOURSELF WEATHER FORECASTING If you know the season of the year, and the direction and speed of the wind, you may be able to predict the weather with far more accuracy than your local weatherman does. How? With a portable weather computer invented by meteorologist Steven Root, of Salt Lake City, Utah.

To use the computerized forecaster, the operator simply enters on the computer keyboard wind direction, wind speed, and the date. A built-in barometer and thermometer provide information on temperature, humidity, and approaching storm fronts. The computer, programmed with the forecasting rules used by the meteorologists in the specific city or town, then displays predictions on a front panel.

According to Root, the computer can predict anything except such meteorological disasters as hurricanes or tornadoes. "Quite frankly," he says, "that's not the type of business I'd like to be in anyway."

Root expects to market his weather computer soon for less than $100.

Onboard mapping computers would warn pilots of bridges and other obstacles by literally drawing them.

MAPS IN FLIGHT The night is stormy and black, and you are flying your fighter plane at treetop height to avoid enemy radar. At 600 miles an hour, you have no fear of crashing into an unseen hill or building. The reason: Your new mapping computer draws pictures of the landscape that you can study as you fly.

This mapping computer is now being built by engineer Louis Tamburino, of the Wright-Patterson Air Force Base, in Dayton, Ohio. As Tamburino explains it, before takeoff pilots simply insert into the computer an electronic memory with information on the terrain of the proposed flight. When the plane is in the air, a navigational computer supplies the altitude, latitude, and longitude to the mapping computer, which rapidly searches its memory for the description of the land below.

Then, using a series of geometrical equations, the computer performs millions of computations to create 30 pictures a second showing the shifting scene outside. The pictures are projected onto a display in front of the cockpit window, and the pilot sees the landscape almost as clearly as he would on a sunny day.

The pictures, formed by white and green contour lines against a background of black, indicate the presence of natural *and* man-made obstacles, including mountains, buildings, and high-tension wires. When a prototype system is completed, says Tamburino, pilots will test the maps in flight.

SIMULATED BASEBALL In just a few years, baseball players may be practicing hitting without stepping onto a field or even using a ball. The sport could be revolutionized by a machine that simulates the experience of batting against a pitcher.

A small-scale prototype of the device is now being developed by Mel Knopf, an Indianapolis aerospace engineer who has worked on the space shuttle and the Apollo project.

To use it, you stand with a bat in your hands over a regulation home plate

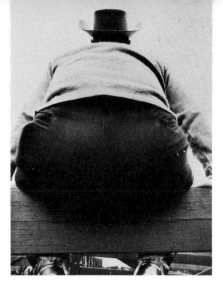

The solution to
this man's obvious problem
may be a fork with
tiny red and green lights.

in a small dark room, facing a sharply curved 200-square-foot movie screen 60 feet away. On the screen is a pitcher—previously filmed by special cameras—winding up and throwing the ball in your direction. The images you see duplicate conditions on the field—a man on second base in the distance, fans behind the baselines—and are conveyed with three-dimensional verisimilitude.

Your bat is coated with a special chemical responsive to sensors. As you swing it, a computerized control console nearby indicates whether you hit the ball with the middle or the end of the bat, swung high or low, early or late. You actually hear a tick if you tip the pitch and a resounding crack if you belt it solidly.

With the simulator batting machine, hitters can work out against the likes of Nolan Ryan and Ron Guidry at any time, slowing down the pitches at will to observe telltale factors such as spin and trajectory.

"What you're doing is fine-tuning the batter so he can hit just about any pitch thrown by any pitcher," says Knopf. "It would provide baseball with an adequate supply of quality hitters. It's so realistic that if a pitcher threw a bean ball, I'll bet the batter would duck."

The machine is designed along the same lines as simulators for training pilots and astronauts for takeoffs, landings, and in-flight situations. It features an electronic optical tracking system patterned after those that monitor airplanes, missiles, and submarines.

Several major league baseball teams are still considering whether to invest the $5 million Knopf estimates it would take to develop the batting machine. Meantime, the engineer is at work on a simulator for pitchers.

DIET FORK A new dieting device now sold in novelty shops is designed with a stop-and-go traffic light for slowing down speedy eaters. Called Slenderfork, it's a battery-powered stainless-steel fork with two tiny lights on the handle. A green signal lasting six seconds gives you the go-ahead to feast

> *"The most beautiful experience
> we can have is the mysterious. It is
> the fundamental emotion that
> stands at the cradle of true art and
> true science."*
> —Albert Einstein

away until a 25-second red one tells you to cut it out.

"It's not a gimmick. It really does work," says Joe Caruso, inventor and manufacturer of the Slenderfork, who slimmed down from 310 pounds to 180 pounds with the unique eating implement. "I was digging a grave with my own fork."

Caruso came up with the idea three years ago, when he experimented with an unorthodox dieting technique—walking around the dining room table between bites. "I said to myself, 'There must be a fork somewhere that jumps out of your hand or electrocutes you for eating too fast.'" Now the former hairstylist even takes the Slenderfork to restaurants with him.

The diet fork is aimed at reforming the behavior of overweight people who eat fast out of habit, since they unwittingly eat too much in the process. Physicians specializing in weight control believe it generally takes 20 minutes for the food you ingest to trip off a "satiety mechanism" in the brain. By eating slower, you notice in advance that your stomach is full, you eat less, and, eventually, you lose weight.

"Of course, the Slenderfork works only if you pay attention to the lights," cautions Caruso, who calls the device an "appestat—like a thermostat for the appetite."

So far, the fork is selling by the thousands, according to Hammacher Schlemmer, a New York gift shop that retails the item. In fact, doctors around the country are prescribing the Slenderfork for dieting patients, many of whom have reported losing weight as a result.

TALKING BRIEFCASES Soon, on those long business trips far from home, you'll have a constant companion to talk to—your briefcase.

Today you can learn to identify radio isotopes by using a talking briefcase developed by EG&G, Inc., in Santa Barbara, California. The briefcase contains a phalanx of radiation-detection equipment and microprocessing units for voice synthesis.

Activity in the buttocks isn't high enough to stimulate blood flow, so a change in driving position is necessary.

As the nuclear components read the output of isotopes, the computer-generated voice keeps up a steady monologue with the user. It tells him what to do next, warns of excess radiation, and reassures the inexperienced tester regularly that the situation is normal.

The briefcase was designed by EG&G to allow nontechnical personnel to identify isotopes and is, they say, just the first of a long line of job-related talking briefcases that will allow users to perform complex tasks at a high level of competency with little or no training.

RIPPLING SEAT One of the biggest discomforts of long-distance driving is the wear and tear it inflicts on the human posterior. Engineers, physicians, and human-factor specialists have labored long and hard to create a comfortable vehicle seat for the long haul, but to no avail.

Now a division of Gulf and Western Manufacturing Company thinks it may have broken through. Engineers at H. Koch & Sons decided that previous failures were due to the act of sitting itself. Muscle activity in the buttocks isn't high enough to stimulate blood flow; long sitting sessions cause blood to pool in the lower trunk, making it sore and tiring the rest of the body, which must work with a diminished blood supply.

So the Koch designers came up with a vehicle seat that ripples. The seat consists of a cushion that is like a bladder divided into small chambers, under which lies a solid-state control unit and a compressed air source. On command from the controller, air from the vehicle's secondary cooling system inflates the cushion chambers one at a time.

The result is a gentle rippling effect as the bladder-cushion expands. In each eight-second cycle, the bladder is fully inflated for two seconds and flat for six. After ten minutes of rippling, the seat settles down. The driver can set it off again by pressing a switch whenever his fanny becomes lethargic.

The company states that the seat will keep a driver sitting in one position at peak alertness for three hours.

Smoke is as dangerous an enemy to firemen as heat and flame because a thick haze leads them to make lethal blunders.

FAST TALK Murray Schiffman is building a successful business out of making ordinary individuals sound like disc jockeys on amphetamines. When you call the office of Schiffman, fifty-five, what you hear sounds like a voice that is running for its life: "Thisismurrayschiffman.Imnothererightnow,butif-you'llleaveamessageI'llreturnyourcallassoonasIcan."

The phone-answering machine is an example of Schiffman's claim to fame: the compressed-speech recorder. This machine derives from the notion that the human brain can take in information a lot faster than the human voice can convey it. Unfortunately, if you take a tape or dictation belt and speed it up to the comprehension level of the brain, the sounds turn into a Donald Duck squeak of the most unintelligible sort.

Schiffman found a way to speed up the tape and eliminate the duck talk. He designed a recorder that, in essence, divides the original tape into thin segments. Since most sounds we make last longer than needed to get across their message, Schiffman's machine cuts each sound back to the nib. It can record as few as half the segments on an original tape and still convey all the verbal meaning.

When the modified tape is played, the voice comes out incredibly fast but totally clear. Already, many 30-second radio commercials are actually Schiffman messages originally recorded at 36 seconds so that more information could be crammed into the time frame.

The major beneficiaries of Schiffman's fast-talking machine will probably be the blind. They will be able to "read" talking books now at twice the speed ever before possible.

SEEING THROUGH SMOKE Many times firefighters die in a blaze because they can't see what's going on around them through the thick smoke and haze, and their officers can see neither them nor the source of the fire. The firefighters have to blunder about until they reduce the smoke level enough to see where they should be heading.

Soon all that may change. The English Electric Valve Company has developed a light, portable camera that creates an image from heat instead of visible light. Using a special wide-angle lens that registers only infrared light—such as is generated by intense flames—the camera can see through the smoke with ease.

Armed with a camera, a firefighter can beam back to his superiors a moving image of what lies around him. They can instantly spot the heart of the fire and see all their troops. The image appears on an ordinary video screen that can either be in a command truck or be carried by a fireman.

With the camera, firefighters may be able to plan better strategies and tame blazes with less danger to themselves.

ARTIFICIAL EAR Deafness may soon be a thing of the past. An artificial ear that uses microelectronics to replace damaged or missing ear parts has been successfully tested in 150 people.

The implanted ear works from a microphone carried on the user's belt. It transmits sound signals to a small electronic grid that is surgically implanted on the inside of the mastoid bone behind the ear. These signals energize a light-emitting diode that shoots a beam of light through a transparent eardrum to a photocell in the middle ear. Here, the light flashes are electronically converted into pulses that stimulate the auditory nerve just as sound waves would.

The most successful implants to date allow people who had been totally deaf to hear about 40 percent of what is said to them. Better artificial ears should be close behind. However, artificial hearing is regarded by some scientists to be even more difficult to perfect than artificial sight.

"While we haven't cracked the code linking the ear to the brain," says Robert L. White, director of Stanford's artificial-ear project, "we'll either have good speech recognition in two years—or never at all. We're running out of excuses."

Teacher works with deaf student. If Stanford is successful, sign language will be replaced with electronics.

NONREVERSING MIRROR David Thomas has proved that playing with soap bubbles can definitely be good for you. Thomas, twenty-nine, was a physics instructor at New Mexico Institute of Mining and Technology when he began fooling around with soap bubbles as part of a project for his students. He concocted a saddle-shaped wire frame and dipped it into the bubbles. When it emerged he saw a reflection of himself in the soap.

Only one problem: His image wasn't reversed, as it would be in a normal mirror. His left hand was on the mirror's left. And that, according to all the physics books, was impossible.

"His achievement," says Kenneth W. Ford, a physicist and president of New Mexico Tech, "is similar to a miner going into an abandoned gold mine and finding gold that everyone else had overlooked."

Apparently, the curved surface of the soap bubbles captures the incoming light and takes it around full circle before releasing it, producing a nonreversed image.

Surprisingly, although Thomas has received a patent on his real-life mirror, he has had little luck finding commercial partners. But he's convinced the applications will emerge over time.

He notes that the mirror can separate individual flashes of lightning and that it can also make rotating objects appear stationary in testing situations. And, perhaps most important, it will keep American men from tying their neckties backward.

LASER SPOTS STOLEN GEMS Shining a laser beam at a diamond produces a unique pattern of scattered light that can be used to identify the gem. Such gemprints have so far been used to identify more than $870,000 worth of stolen diamonds. The evidence is so damning in court, says Craig Carnevale, general manager of Gemprint Ltd., in Chicago, that the first ten defendants to come to trial all pleaded guilty.

Because diamonds are individually cut, no two stones have exactly the

215

Each gem has its
own unique "fingerprint,"
which can be
recorded and detected by lasers.

same pattern of facets. Each diamond's unique facet pattern is recorded in the gemprint, a Polaroid photograph of laser light reflected by the diamond. The patented technique was developed at Israel's Weizmann Institute of Science. The equipment is made by Kulso Ltd., in Haifa, Israel, and sold in the United States by Gemprint Ltd.; a typical system sells for about $3,500.

Carnevale says that several hundred jewelers in the United States use gemprint systems, as do the Federal Bureau of Investigation and police in Chicago, Los Angeles, Dallas, Denver, and Miami. One copy of each gemprint goes to a central registry in Chicago, where the image is automatically translated into computer-compatible form and stored in a computer file along with identification of the owner. When a stolen diamond is recovered, police can submit a gemprint to the registry for computerized comparison with others on file.

So far over 100,000 gemprints have been recorded. Among the stones registered are the Hope diamond and other large diamonds in the Smithsonian Institution's gem collection.

Carnevale cites testimonials from law-enforcement officials who have used gemprints to identify diamonds and convict the thieves. But some of the most impressive testimonials are from insurance companies, concerned because normally less than 3 percent of all lost or stolen diamonds are returned to their rightful owners. Commercial Union Insurance of Boston will pay its policyholders to make gemprints of diamonds insured for more than $5,000, and several other insurance companies offer 10 percent discounts on premiums for gemprinted diamonds.

ENVIRONMENT

CHAPTER 9

Survivalists: ". . . not a threat to anyone. What they're trying to do is preserve society and its laws."

SOVIET CHILL Whenever it's too hot or too cold, there are those who blame the Russians. A hundred years from now there could well be cause for complaint. According to a scientist at Western Michigan University, Soviet plans to divert some rivers in the USSR could lower the temperature of the Northern Hemisphere and drastically alter climate around the world.

Geographer Philip Micklin reached this conclusion with the help of a flow chart that analyzes Soviet plans to divert water from the Ob and Yenisey rivers in the North to agricultural and industrial regions in the South. The rivers (each about as long as the Mississippi) now flow to the Kara Sea, which in turn empties into the Arctic Ocean. Reducing the flow of water to the Kara, and then the Arctic, Micklin explains, would make it more difficult for the North Polar ice cap to melt. The results: a larger ice cap and a colder Arctic Ocean. "It is a logical conclusion," he says, "that a colder Arctic would cause a generally cooler Northern Hemisphere and certainly more variable weather."

None of this will happen overnight. Indeed, the initial stages of the Soviet plan call for the diversion of a mere 60 cubic kilometers of water a year, which will not have much impact on world environment. But, Micklin warns, the long-range schemes are far more grandiose. "The Russians are well aware of the dangers," he notes, "but it might be almost impossible for them to resist the tremendous political and economic advantage of getting more water to the South."

SURVIVAL CONDO People determined to survive anything from nuclear war to social collapse in recent years have created a market for everything from crossbows to radiation suits. Now they can survive in style—if they have $80,000 to plunk down for an underground condominium.

Survive Tomorrow, Inc., a Utah-based firm, has built an underground condo project in the Utah town of LaVerne (population 1,200). Located far from cities or military installations and protected by mountains, the condos are designed to withstand anything except a direct nuclear hit. Above each condo are three feet of earth and eight inches of reinforced concrete. An independent utility system, safeguarded air- and water-filtration systems, a decontamination center, and medical facilities are part of the package.

While the base price of each unit is $39,000, each comes with a four-year supply of freeze-dried and air-dried food for four persons, with storage space above the shower and in other out-of-the-way spots. This raises the price of the condos to nearly $80,000.

STI president Ron Boutwell says, "Unlike gun nuts and other kooks, organized survivalists are not a threat to anyone. What they're trying to do is preserve society and its laws."

MISSISSIPPI SEA About 1 billion years ago a lengthy rift began to form in the middle of North America, from the Gulf of Mexico to southern Illinois. This Mississippi Valley rift might have split the continent in half had it remained active, and it may yet do so.

Two Northern Illinois University geophysicists, Lyle McGinnis and Patrick Ervin, have advised industries planning to build near the rift to design structures able to withstand an earthquake that would register eight on the Richter scale. McGinnis says, "Twenty earthquakes ranging from magnitude one to two occur along the boundaries of the rift every month. Occasionally they reach magnitude four or five. In the early 1880s a series of major earthquakes, including the largest ever recorded in North America, occurred along this fault."

When the last major quakes occurred along the rift, the area was lightly populated, but now "one would be as devastating as any quake that ever struck California," McGinnis says. Furthermore, a large quake in the center of the country would radiate its energy to a greater distance than those in California, probably causing damage over a much wider area.

Though the rift is now in a relatively dormant stage, such faults have a

tendency to become reactivated. "Conceivably, the continent could even open up to form an inland sea," Ervin suggests, though he is quick to add that this could take hundreds of millions of years.

BURMESE ECO-ENGINEERING While environmental-action groups in America langorously debate the importance of ecological engineering and appropriate technology, peasants in central Burma are proving that these concepts work.

Inle Lake, in the heart of Shan State, the mountainous heartland of Burma, contains an isolated society based upon human ecological engineering and the development of locally appropriate technologies by the indigenous Intha tribes.

Because there is little arable land in the region, the Intha have always focused their lives on the lake. But, unlike other lakeside Asian people, they don't live in wharfside boats; instead, they've created their own islands. When silt began forming around the stilts that used to hold up Intha houses by the lakeshore, they aided the process by slapping on mud and weeds to form initial ground cover to anchor their islands.

In their livelihood, too, the Intha are unique. They don't fish their lake, they farm it. Today, there are 50 acres of tomato beds floating on the 12-by-14-mile lake. The plants are anchored by mud and weeds to poles driven into the lake bottom. Man-made reefs protect the beds from waves and weather.

The result is one of the most prosperous rural societies in Asia. It's self-sufficient, nonpolluting, and highly imaginative. And it didn't require a Department of Energy grant.

SPOILED PARKS What has traffic jams and nightclubs, a full-time hospital with three doctors and a dentist, and a luxury hotel that ranks with the finest in the world? Answer: Yosemite National Park, in California.

According to Armond Sansum, an official at Yosemite, our national parks

Spoiled parks:
If current conditions persist,
they could result in
"permanent ecological destruction."

are being destroyed by money-making concessionaires who operate everything from gift shops in the Grand Canyon to golf courses in Yellowstone. In order to be profitable, says Sansum, these businesses require a heavy stream of customers. To maintain that stream, they create commercial attractions by the score.

Apparently, their strategy has worked. In 1980 national parks were visited by 238 million people, nearly twice as many as a decade ago. Yosemite alone ministered to 15,000 people every day of the summer. Its campgrounds were lined with wall-to-wall tents, and roads were choked with traffic, including trucks that continually brought in food and took out garbage. A new $6 million sewage system proved too small to meet the state's quality-control standards.

Right now, says Sansum, the damage caused by industry and overcrowding is reversible. But if current conditions persist, they will result in permanent ecological destruction of the parks. The only solution, he adds, is removing all the luxuries. Until that happens, people who want to enjoy natural splendors such as forests, waterfalls, and meadows will be disappointed. Indeed, tourists who visit parks today are likely to find the same "urban" problems they had hoped to leave behind. One recent Yosemite crime report listed 2 manslaughters, 5 rapes, 36 assaults, 622 robberies, and 465 traffic accidents—all in a single day.

POISON PARSNIPS The perils of red dye Number 2, cyclamates, nitrates, and other man-made chemicals found in foods are well documented. But these may pale beside the toxins that occur naturally in certain plants and vegetables.

A newly discovered source of possibly harmful ingredients: parsnips. The vegetable contains a trio of chemicals, called psoralens, that can trigger skin cancer in animals when they're exposed to sunlight and that are toxic to mammals.

Some clever reptiles,
enlisted by the U.S. Fish and
Wildlife Service,
surrounded unsuspecting suspects.

So far the only humans known to have suffered ill effects from psoralens in vegetables are the pickers and processors who handle the plants. A photo-induced dermatitis sometimes ensues. So potent is their effect on the skin, doctors use psoralens to treat vitiligo (skin depigmentation) and psoriasis.

Scientists at the U.S. Department of Agriculture's Veterinary Toxicology and Entomology Research Lab, College Station, Texas, tried stewing the parsnips to find out whether cooking kills off the psoralens. Neither boiling the roots nor zapping them in a microwave oven destroys the chemicals. But peeling the parsnips did reduce the amount of psoralens.

Dr. Wayne Ivie, of the College Station research lab, says that the study doesn't suggest that parsnip lovers should give up the vegetable. To date there's no indication that humans become sick from eating the roots.

"Mankind is blessed with very effective mechanisms for dealing with toxic chemicals," contends Ivie, who will continue studying psoralens occurring in both human and animal foods.

"That the body has the ability to rapidly metabolize and eliminate these toxins should be reassuring and should make people feel safer about being exposed to other potential poisons, such as pesticides."

SNAKESCAM Covert "sting" operations are the rage among federal law-enforcement agencies these days. Even the U.S. Fish and Wildlife Service has launched an undercover investigation called Snakescam. The two-year plot to trap profiteers who sell endangered reptiles on the black market led to the biggest bust ever of wildlife smugglers in the United States. According to special agent Rick Leach, 27 suspects in 14 states were arrested.

Fish and Wildlife Service agents borrowed techniques from other law-enforcement experts who stalk drug smugglers and crooked congressmen, Leach says. Operating through a phony storefront business called the Atlanta Wildlife Exchange, in Atlanta, Georgia, they arranged and taped transactions, then made arrests.

Air-dropping pesticides can help save corn and apples from insects, but they're also killing valuable honey bees.

In a year and a half, Leach says, the agency bought and sold nearly 10,000 live animals illegally plucked from the wild. Many were rare and exotic reptiles classified as endangered species: the Indian python, for instance, the American alligator, and even the Gila monster, a small, venomous lizard. Indeed, the investigators discovered that native American species fed a black-market trade that flourished to the tune of up to $100 million a year, including a booming business in Europe and Japan, where collectors pay many times the domestic price for a prized specimen.

Penalties for those arrested have included fines of up to $7,000, five years' probation, and compulsory terms of public service that may involve work in wildlife conservation, Leach says.

HONEY CRISIS It has become a real problem in New York state. Bees are dying off by the thousands because of pesticides meant for other insects. "We found that in 1980 alone, we had nearly 5,000 colonies of bees affected by pesticide use in the state," complains Roger Morse, the head of the bee research at Cornell University, in Ithaca, New York. "And it doesn't seem to be getting any better."

Morse says that he checked the honey production in a number of apiaries and found that it was off by a significant amount. Part of the decline was due to the poor weather conditions during the past year. But part of the lack of honey production was due to inadvertent poisoning.

Pesticides are used to check the populations of several insects. In New York, where sweet corn and apples are major crops, pesticide use is crucial. Unfortunately, bees in the area are not spared the effects of the poisons. Pesticides have also been used extensively to curb the populations of gypsy moths that have swept through the Northeast. One of the major chemicals used to control these moths, Sevin, also kills bees.

These preventive measures won't ease up in the near future. "There is a new pesticide on the market that's been giving us trouble," says Morse.

Fluorescent lights:
We're just becoming aware of the
biologic implications
of our lighting environment.

"The gypsy moth is increasing its range and is a greater problem than it has been. So these two things coupled together aren't helping matters any."

Morse believes that the solution to the problem is to talk to the growers. "There are certain pesticides," he says, "that are safer than others. There are also methods of application that make a big difference." He plans to make certain recommendations to all of the parties concerned and to hope for the best. "It's a question of education. The people using these materials have to know what happens to the area when they spray."

DANGEROUS FLUORESCENTS Fluorescent lights: They're energy-efficient, cool, and relatively inexpensive. But they may also be hazardous to your health.

Colin Chignell and others at the National Institute of Environmental Health Sciences, in North Carolina, have found that mice who live a large part of their lives under fluorescent lights show stunted growth, smaller numbers of offspring, shorter lifespans, and more of them develop cases of mammary tumors.

The mice, which develop mammary cancer naturally, were exposed to the artificial light 12 hours a day for more than a year and a half. The results were alarming. Most of the mice raised under the cool, white lights were worse off than those raised under daylight-simulating ones.

Chignell's study, unfortunately, is not alone in its findings. Others have discovered similar problems with artificial lighting. Most agree that the light spectra are to blame. The spectra, or array of energies in light, of artificial lights are often markedly different from those of sunlight. The spectra of many kinds of fluorescent lights differ to an even greater degree.

No one yet knows how these artificial lights affect body chemistry. "Those are questions for which we don't have answers," Chignell says. "All we have is observed effect."

Other researchers are more nervous about the possible subtle effects arti-

Do you feel the air
is fresher in the forest?
Trees, say experts,
help filter out pollutants.

ficial lights have on our bodies. "For the past few generations," writes Richard Wurtman, a professor of endocrinology and metabolism at the Massachusetts Institute of Technology, in *Neuroendocrinology*, "people have spent much of their lives under artificial light sources, designed to deliver the greatest intensity of light most economically, whose spectra bear little similarity to natural sunlight under which life on earth has evolved. We are now just becoming cognizant of the possible biologic implications of tampering with our lighting environment."

TREE FILTERS Tired of the car fumes and the dust? Stand under a tree. It might just make breathing a little cleaner. That's the opinion of William Smith, of the Yale School of Forestry and Environmental Studies, in Connecticut, who has found that trees make excellent air filters.

Smith walked around downtown New Haven and took samples of the London Plane trees planted there. Back in his laboratory, he analyzed their leaves by measuring how much light they absorb and what their surfaces look like. Scanning the leaves with an electron microscope, he found everything from pollen grains and bacteria to dust and trace metals on the leaf surfaces and embedded within.

Trees, he believes, act like huge passive filters by exposing their leaf surfaces to the atmosphere. These large surface areas help to remove particulates from the polluted air. "If you look at the surface volume ratio of a large tree," he explains, "it's absolutely enormous. So by virtue of presenting that surface area to the atmosphere, they have the opportunity to accumulate particles from the air."

Surface area is not the only factor involved in the tree's ability. Adds Smith: "It's not the same thing to present a plywood building to an airspace as it is to present something as geometrically complex as a tree. There's much more surface and opportunity there for retention."

So should people start planting more trees in polluted cities? Smith

"Man is a complex being. He makes deserts bloom—and lakes die."
                                        —Anonymous

doesn't think so. Too many would be needed. But, he adds, forests may be more than a nice place to spend a vacation. "If you look at the studies that have been done, the relative amount of particulates in the atmosphere that's within a forest area and outside a forest area—two adjacent areas, same particulate loading—the forest is invariably cleaner, at ground level."

RADIATION SHIELD It sounded like the perfect thing to put into the field packs of soldiers who might have to face the effects of radiation in atomic warfare. But the drug, WR-2721, was found to be too impractical by the Department of Defense when it learned that the agent wouldn't work if taken orally. That's when a tumor biologist, now working at the University of Pennsylvania, decided to see whether it could be used in cancer therapy.

The reason for John Yuhas's decision was the observed effects of WR-2721 on animals. Mice injected with the substance could tolerate 40 to 100 percent more radiation to their organs and 200 percent more to bone marrow than mice not given the injections.

The drug protects most tissues from the toxic effects of radiation by binding to the highly excited atoms in the cells that are stimulated by the radiation. Yet WR-2721 does not shield cancerous tumor cells, because of their cell membranes. The cellular "skin" has the ability to keep out such water-soluble drugs. Eventually, the radioprotector is flushed from the body.

So far, the research is only in the beginning phases. "We're trying to find out what is the maximum dose that the patients can tolerate," says Yuhas. The tests are conducted daily along with radiation therapy and at longer intervals with chemotherapy. But, eventually, the cancer therapy doses will be gradually increased.

Even though the treatments are currently being used on cancer patients, Yuhas believes that this is not the only use for WR-2721. "There is no reason a normal person would not respond to the drug, as well or better than cancer patients." The drug is already being used to treat patients with cystic fibro-

A drug originally meant to protect soldiers from radiation may eventually be used on cancer patients.

sis, a disease of the lung. "It's in no way curative," he cautions. "It's just a method of giving relief from this killing disease."

Thus far, WR-2721 is the only radioprotector that has been approved by the FDA for clinical testing. But if the testing is successful, it could have wide-ranging uses not only in cancer therapy but in high-radiation zones as well.

RUSSIAN INVADERS The State Department may not know it yet, but there are Russian plants in New York. These plants, however, have green leaves and roots instead of the usual trench coats and code books.

The plants, called giant hogweeds, were originally grown in the Soviet Union and were brought into the United States as conversation pieces. They have since escaped the confines of their owners' lawns and have spread throughout most of the central and western parts of New York state.

What makes them dangerous is their ability to inflict anyone who is unfortunate enough to brush up against one of these things with severe rashes and scars that last for years. The weeds, which grow along stream banks and waste areas where water collects, produce skin irritations known as phyto-photo-dermatitis. The irritations occur in the presence of moisture and sunlight some 24 to 48 hours after contact with the plant. The victim quickly develops rashes on the affected areas; these are followed by swelling and blistering. But unlike poison ivy rashes, once these rashes are gone, scars and discoloration stay around, often for years. Fortunately, anyone passing near these weeds can usually recognize them by their Olympian stature, meter-long leaves, and four-meter breadth. The five-meter tall plants, which are related to parsnips, also produce showy dome-shaped clusters of white flowers and seeds that look like they come from the sunflower plant.

So far, say the experts, the hogweed hasn't ventured into other states. But if you happen to find one of these giant weeds on your lawn one day, carefully destroy the entire plant, including the root. Botanists warn people that if they don't do this, the hogweed will just shoot up again the following spring.

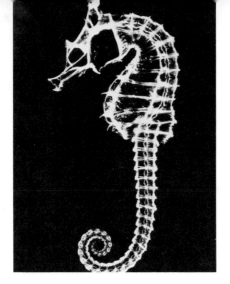

Profiteers are
blasting coral reefs to bits
in the course of
capturing tropical fish.

TROPICAL RIP-OFF Walk into any school, and you see tanks filled with dazzling coral and dozens of clownfish or moorish idols. You probably don't realize, however, that the people who supply these aquariums systematically destroy delicate coral reefs, then ship the captured fish with little or no care for their survival.

According to British ecologist Tony Loftas, half of all fish captured in tropical waters die before reaching a hobbyist's tank. The reason? Exotic-fish companies spring up for several months or a year, make a quick buck, then go out of business. Because they plan on staying in business for only a short time, these firms don't bother preserving the marine habitat or protecting the fish. At any one time, these get-rich-quick pirates make up half of the tropical-fish trade.

Tropical-fish traders have already wiped out entire fish populations in South America, Sri Lanka, and Africa. In coastal areas of Kenya, Indonesia, and Florida, coral reefs are being blasted to bits so that the pieces can be used to decorate aquariums. And poisons used to stun certain fish slaughter other, nontarget animals.

Some people suggest that strong international regulations could put a halt to the destruction of the marine habitat. But the regulations are hard to push through because importers have an enormous influence on the U.S. Congress and other governmental bodies. Moreover, such regulations could triple the cost of fish for millions of hobbyists. For now, ecologists say, the best solution may be educating hobbyists to take a greater interest in where and how their fish are collected.

MONSOON MADNESS While studying the water levels of lakes in arid parts of Africa and Asia, geologists discovered something they couldn't explain: The levels have been dropping for 9,000 years.

A meteorologist from Wisconsin may have the answer. John Kutzbach, a professor at the University of Wisconsin-Madison, believes that changes in

> *"Scientists have odious*
> *manners, except when you prop*
> *up their theory; then you can*
> *borrow money of them."*
> —Mark Twain

the earth's orbit have affected the intensity of monsoon rains.

In past geologic ages, monsoons sweeping across the Sahara and Asia packed an extra wallop. Why?

Using a model of the atmosphere to simulate the climate 9,000 years ago, Kutzbach calculated that the earth passed closest to the sun in July (in the Northern Hemisphere). This means that summers then were 7 percent hotter than now; winters 7 percent colder than today. (Today, the earth is *farther* from the sun in summer, closer in winter.)

This greater contrast in temperature resulted in stronger sea breezes. When cool ocean air moved more readily across rapidly heating land, heavier rains fell.

If weather changes with the earth's orbit, Kutzbach thinks, the Sahara and parts of Asia could see more intense monsoons again. The planet may eventually return to the points in its orbit in which it lies closest to the sun in summer and farthest away in winter.

In the meantime, Kutzbach's findings will give meteorologists a better understanding of weather conditions occurring today.

POLLUTION CANARY Old-time coal miners used to bring canaries into their mines to warn them of coal gas. If the deadly stuff seeped into their area, the tiny bird would go first; when he stopped singing, they got out of there.

Modern workers exposed to radiation have a high-tech canary in the form of a badge that darkens with radiation exposure. But modern miners and those who work with other organic pollutants have to use a clumsy, expensive warning system that is little more than a mechanical canary. Air is sucked past a polymer filter that absorbs dangerous compounds. At intervals, the filter is heated and the result is read by a gas chromatograph mass spectrometer.

Workers dislike the pumps because they hum. Also, unless the pump is

The itinerant rainmaker may be no less efficient than our meteorological experts.

right next to the workers it may give a false reading. Now, the Environmental Protection Agency (EPA) is working on a danger-detection badge for organic pollutants. Monsanto has made a prototype badge that tested well against the pump, but technical problems remain. EPA hopes eventually to have a badge that recognizes nine common pollutants.

Even if the badges merely match the pump in sensitivity, they will offer better protection at lower cost, says James Mulik of EPA. "You could put out a thousand of these badges a lot cheaper than you can a lot of sophisticated instruments. People don't mind wearing a badge, but they just don't like wearing pumps."

IT'S STILL HARD TO CHANGE THE WEATHER People have tried to make it rain since prehistoric times, and atmospheric scientists have been working on the problem for the past three decades. In 1977 what's now called weather modification—including stimulating snowfall and stopping hailstorms—was tried over some 7 percent of the land area of the United States. However, it's still a lot easier to talk about the weather than to do something to change it.

After reviewing past research, the National Research Council's Committee on Atmospheric Sciences concluded that: "The effects of weather modification efforts . . . have yet to be definitely established." The only exception is clearing fog that has been cooled to such a low temperature that it's no longer stable. The committee's report adds: "The natural variations of precipitation are so great that it is difficult to detect the difference cloud seeding may have made in any given situation."

The panel does cite some encouraging results, including "limited evidence that suggests it might be possible to reduce the high wind speeds in hurricanes." However, it recommends that near-future efforts concentrate on improving our understanding of the fundamental processes that cause precipitation. There's a long way to go before we can tame the weather.

THE
PARANORMAL
(AND OTHER WEIRD THINGS)

CHAPTER 10

> *"My figures coincide in fixing*
> *1950 as the year when*
> *the world must go to smash."*
> *—Henry Adams*

HUMAN COMBUSTION Jack Angel, a traveling salesman on business in Savannah, Georgia, reportedly burst into flames while asleep in his mobile trailer. Believed by some to be the only survivor of spontaneous human combustion, Angel was left with a hole in his chest, fused vertebrae, and an arm so charred it had to be amputated.

According to David Fern, a Savannah physician summoned to the scene of Angel's accident, his patient was obviously a victim of spontaneous combustion (a bizarre molecular reaction, Dr. Fern says, that causes people to burn up inside). "There is no other plausible explanation," he contends, "because objects in the trailer were not the least bit singed. Thus, they could not have set Jack Angel on fire."

Others disagree. Dr. Marion Jordan, director of the burn unit at Washington Hospital Center, in Washington, D.C., for one, says that Angel may have taken an antibiotic containing sulfur. One adverse reaction to such a drug is shrinkage of skin and blood vessels, creating the appearance of a true burn.

Before his experience Angel earned $70,000 a year and was in peak physical condition. Now collecting Social Security checks, he is confined to his home. Though he has no memory of the incident, Angel clings to the belief that his injuries resulted from spontaneous combustion.

DOGU SPACE SUITS Dogus are small clay statues with pointy heads, insect eyes, and torsos marked by intricate patterns of dots and stripes. They were made in Japan, between 7000 B.C. and 520 B.C. Some people think they represent Japanese fertility gods. But, according to Vaughn Greene, author of the book *Astronauts of Ancient Japan*, these artifacts actually depict space-suit-clad visitors from another planet. The most striking evidence to date, Greene says, is the similarity between dogu markings and the new National Aeronautics and Space Administration (NASA) space suit—the extravehicular mobility unit (EMU) to be worn by space shuttle astronauts outside their ship.

According to the latest census, Massachusetts is the state of choice for American vampires.

For instance, Greene says, the chest-pack control units on the EMU are in roughly the same place as circular knobs on a dogu chest. These knobs probably controlled life-support systems on the dogu space suit, he asserts, just as they do on the EMU. And the stripes surrounding the dogu knobs are simply markers to calibrate the quantity of water or oxygen being dispensed to the person in the space suit. From a bulge in the dogu's midriff, Green theorizes that the top and bottom of the dogu space suit were put on separately, just as the EMU is.

One NASA scientist notes that a highly advanced, humanlike civilization would probably design space suits far more sophisticated than those Greene says are found on the dogu. Greene suggests that if our current space suit could get us to the moon, it could also get us to another planet.

FANG COUNT Dr. Stephen Kaplan, of the Vampire Research Center, in New York City, has completed the world's first vampire census. What were the results? People in Massachusetts should worry.

After receiving more than 500 responses worldwide, Dr. Kaplan found 21 "physical vampires"—defined as those who "survive by drinking human blood"—in the United States. For some reason, he says, U.S. vampires like Massachusetts best, with Arizona, California, New Jersey, and Virginia following in popularity. In addition, he received scattered reports of cases in Canada and in countries as distant as Japan and Germany. Those counted in the census ranged from fifteen to forty-one years of "apparent" age, but some claimed to be as much as three hundred years old.

Dr. Kaplan, a self-styled vampirologist, had appeared on hundreds of radio and television shows, asking for information "from people who know vampires, or have vampiric tendencies, or claim to be vampires.

"This is no spoof," he insists. "Any myth or legend may have some basis in reality, and we wanted to find out what the reality is in the specific case of vampires.

233

"There are uncertain areas," he admits. "Blood cultists, fetishists, the kooks, and the just-plain-crazies responded. This is merely part of the fallout one gets in this field."

Some of the vampires who responded to the census do sleep in coffins, but not underground. Kaplan himself asserts, "I never met a vampire I disliked. They're all fascinating . . . as long as they don't sup on my blood."

LOCH NESS WORMS For nearly 50 years investigators from both sides of the Atlantic have been lurking about Scotland's Loch Ness in search of a huge monster nicknamed Nessie. Thus far they've had no luck. But now three zoologists studying the Loch Ness phenomenon have discovered a monstrous welter of American flatworms, insects never before seen in Europe. The pressing question is, How did the flatworms get there?

At first the University of North Wales scientists suggested that worm cocoons, clinging like limpets to the outsides of ships, were dropped into the Caledonian Canal, of which the loch is a part. This notion was eventually dismissed, however, since there is no way that cocoons could survive the ferocity of the ocean all the way from America to Scotland.

Instead, it has been deduced that the worms' presence is due to none other than Nessie, the monster, or, more precisely, the search for this aggravatingly elusive creature. The zoologists think the worms, which have the knack of sneaking into narrow crevices and laying eggs on solid objects, arrived in Scotland on American monster-hunting equipment. The Loch Ness worms may not be as formidable as the Loch Ness monster, but so far at least they have more substance.

NOSTRADAMUS INTERPRETED The cryptic prophecies of the astrologer-mystic Nostradamus confounded interpreters in France for five centuries. To escape the menacing eye of the Catholic Inquisition, Nostradamus, a Christian who claimed to be inspired by God, employed a baffling array of

Loch Ness: For nearly
50 years researchers have looked
for a monster,
and now they've found . . . worms?

rhetorical stratagems that disguised his predictions and ensured their perpetuation. Throughout the years his verses sparked some 400 works of interpretation, none of which apparently broke the code.

Then entered the computer. Jean Charles de Fontbrune, a pharmaceuticals manager who had picked up the hobby of "Nostradamizing" from his father, fed the fruits of his years of labor into a national computer network. Thus, he was able to measure the repetition of letters, words, phrases, and other key linguistic devices.

De Fontbrune realized that Nostradamus thought in Latin structures and wrote those structures directly into French. He also plumbed Latin poets, for example, Vergil, for countless word-play techniques, such as anagrams and aphaeresis (dropping the initial letter or syllable from a word).

Soon De Fontbrune had solved 600 of Nostradamus's 1,100 verses; De Fontbrune published a book in 1980 of his findings, but at first nobody much noticed. Then chaos broke loose. Nostradamus, De Fontbrune's book showed, had predicted that "the year the Rose flourished" would coincide with an uprising of Muslims against the Western powers. When the Socialists (whose symbol was the rose) took power in France and when the U.S. embassy in Tehran was seized, the eyes of Paris hastily turned to De Fontbrune's pages. Readers soon realized that Nostradamus had prophesied the death of Henry II in a tournament, the rise of Napoleon, and the overthrow of the shah of Iran by "religious zealots."

The book caused a panic in Paris; some people even pulled up roots and left. According to De Fontbrune's interpretation, before this century ends, Islam will destroy the Roman Catholic Church. Then the Arab world will team with the USSR and invade Western Europe. Paris will swim with blood, and the world will be plunged into a terrible war.

De Fontbrune says that Nostradamus's predictions were intended only as a warning, however, and if all the nations simply shake hands, the cataclysm will not occur.

Haunted house:
Don't worry about that ghost,
it's probably
just some old forgotten trauma.

HAUNTED PERSONALITIES Haunted houses come into existence through the power of the human mind, says D. Scott Rogo, a Los Angeles researcher who's written several books on the subject. According to him, the personality of the main witness in many haunted houses is "basically hysterical, traumatized, and rather repressed. The repressed hostility erupts in a haunting, somewhat the way a kettle gives off steam," he says, and the result is mysterious poundings, flying objects, or spontaneously ignited flames.

Sometimes an act of violence, such as a murder, will set up the physic conditions that provoke strange sounds and visions, Rogo asserts. "When you talk to people in haunted houses, you often find there's a history of trauma in their lives. There do seem to be haunting-prone individuals."

Rogo, thirty-one, says that he lived in a haunted house in Los Angeles in the early 1970s. "I began to realize that you could feel when something was going to happen," he explains. "The house would act up and then die down for a few months. There were a lot of things I would previously have considered fantastical, but I knew they were real when I experienced them first-hand."

Rogo says his latest research indicates that inner human conflicts may account also for UFO abductions and putative apparitions. People who see bleeding statues, he remarks, have probably led traumatic lives.

CHINA'S ARMPIT SAVANTS Are Wang Qiang, twelve, and Wang Bin, ten, able to use ESP to "read" the messages scribbled on bits of paper tucked under their armpits? Absolutely, according to a report in *Nature Journal*, China's prestigious science magazine. Indeed, the publication arranged a conference so that the sisters could demonstrate their powers: Performing before a scientific audience in July 1981, they seemed to describe the messages that were stuffed in their armpits, their hands, and even their ears, with uncanny accuracy.

> *"Kids get their ideas about UFO's*
> *where I learned about sex:*
> *the tabloids and the sleazy press."*
> *—J. Allen Hynek*

The summer seminar piqued the interest of Luo Dongsu, a Chinese Air Force physician who spent four additional months testing the girls. According to the American publication *Parapsychology Review*, Luo took electrical measurements of Qiang's hands and ears and found "sensitivities vastly greater than [those associated with] modern military radar." Finally he suggested that the sisters' "special functions" were due to "the relation between electromagnetic waves and the human body."

Actually, Western magicians have been reading with their armpits for centuries. Their technique is simple: They put one scrap of paper under an arm and hold another scrap in their hand. Wearing loose goggles instead of a tight blindfold, they discreetly peek down to read the message tucked in the palm, then deftly exchange it with the one in the armpit. Because they submit the scrap they have really looked at for examination, these performers always seem psychic.

Like these professional magicians, the Wang sisters might be cheating. According to Cyrus Lee, head of the psychology department at Edinboro State College, in Pennsylvania, the Chinese experiments included several scraps of paper handled freely by the girls, who used gogglelike blindfolds.

REMEMBERING BIRTH In our mother's womb we were kept warm and secure. No pain, no fear, just peace and quiet, and a comforting heartbeat to soothe us. Then, suddenly, we were thrust into a new environment. Noise, light, pain, terror, all came crashing down on us at once. To say the least, being born was traumatic. Why then don't we remember that event?

"The fact is," says clinical psychologist David Chamberlain, "we do." Chamberlain, of the Anxiety Treatment Center, in San Diego, hypnotized several children and found they remembered their birth in vivid detail. Many recalled their mother's hairstyle, the surgical instruments used, the conversations among the hospital attendants, and even the mother's physical and emotional state.

Firewalker: One American
scientist thought he had figured
out the trick—
and ended up in the hospital.

In most cases, according to Chamberlain, descriptions given by both mother and child were nearly identical. "The accuracy of recall," he says, "suggests sophisticated mental activity from the beginning of life, including the ability to store and retrieve memories, to experience, learn, and even understand."

Most other obstetricians and psychologists say hogwash. They attribute these "memories" to our fantasy, explaining that infants simply do not have the intellectual ability to remember anything occurring so early. Nevertheless, Chamberlain is adamant: His research has convinced him, he says, that "memory probably can go all the way back to the womb and may even go back as far as conception."

SOLE ON FIRE When neurologist Christo Xenakis, of Athens General Hospital, studied firewalkers in Langadhas, Greece, he assumed that prolonged contact between human feet and scorching coals would cause third-degree burns. But physical examination of the firewalkers, he says, "proved that exposure of less than one second produced only small blisters on the sole of the foot." After monitoring the Langadhas firewalkers with thermometers and film, the neurologist concluded that "no deception was involved in their ritual; they used their minds," he says, "to combat the feeling of pain."

Xenakis's claims were recently bolstered by a vacationing American scientist who tested the coals himself. "I wanted to know whether it was just a case of having the nerve to jump on the coals and keep the feet moving fast, or whether there was something paranormal about the experience," George Mills says. "I figured that with the right psychological attitude I could beat it and get off with merely a few minor burns." Mills, like Xenakis, was unable to fathom the firewalkers' secret. He is now convalescing from third-degree burns on both of his feet, and adding insult to injury, the firewalkers have accused him of profaning their ceremonies.

Nevertheless, skeptics continue to offer scientific explanations for the practice, including hallucination, pain-inhibiting drugs, and the use of ash to insulate the coals. According to magician James Randi, who investigated the ritual thoroughly, the successful firewalker uses "coals so hot they emit a blanket of dry steam to protect the feet." If people walk over the coals rapidly, "taking no more than five steps in all," he says, "they'll suffer virtually no injuries." The essential thing, he warns, is for firewalkers to keep their feet completely perspiration-free. Otherwise, the dry steam will get damp and scorch the soles.

UFO DETECTORS According to some avid flying-saucer watchers, the well-equipped home should have not only smoke detectors and a burglar alarm system but also a UFO detector. Although do-it-yourself plans can be purchased at a reasonable price, most UFO-conscious buyers prefer the ready-made models.

Shields Enterprises, in Emmaus, Pennsylvania, for example, sells more than 300 detectors annually. Designed to detect the strong electromagnetic fields often associated with UFOs, the Shields device consists of a magnetic reed switcher, a buzzer, and a battery, in a metal container. To activate it, users place the box on a window ledge inside the home and position an accompanying detector probe outside the window. When electromagnetic energy is high, the alarm sounds. In addition, the battery and detector can be tested by depressing a red button on the front panel.

Some ufologists protest that since not all UFOs demonstrate electromagnetic properties, overconfident detector owners might waste their nights in slumber while fleets of extraterrestrial spaceships pass over their rooftops. Critics also claim that most magnetic detectors fall into two categories: those that remain obstinately silent because they are unresponsive to any but the strongest magnetic fields, and those that are oversensitive, sending hopeful UFO observers rushing to their windows at all hours of the day and night.

Despite the controversy, consumers continue to buy the devices. The

Meditator: Can a
TM field of consciousness
permeate a
neighborhood and reduce crime?

Shields UFO detector, which sells for $17.95, comes with a 100 percent money-back guarantee. The manufacturer claims it has never received a request for a refund.

OM SWAT TEAM In Atlanta gentle, mantra-chanting Transcendental Meditators (TMers) are serving as a sort of astral auxiliary police force.

A group of 28 to 40 TMers, meditating together nightly for two weeks, reduced violent crime—murder, rape, and aggravated assault—by 20 percent in Atlanta's crime-ridden Grant Park section, according to social psychologists Elaine and Arthur Aron. Felonies reportedly jumped back to normal levels as soon as the meditators returned to their center in an affluent neighborhood.

While crime rates commonly fluctuate, Arthur Aron points out, "the probability is almost nil that these results happened by chance." So what accounts for the "Maharishi effect" (named after TM founder and guru Maharishi Mahesh Yogi)?

Well, just as quantum mechanics experiments reveal the existence of a "nonphysical field," the unseen "field of consciousness" affected even nonmeditators within its range, Aron says. In Atlanta, he reasons, the "neural coherence" generated by the TM practitioners permeated a neighborhood, inhabited by 70,000 people, with "pure consciousness," producing "measurable social effects."

All of the Atlanta meditators, like the Arons themselves, were graduates of the TM Siddhi courses, which teach meditation-propelled "flying" and other spiritual powers *(siddhis)* like "knowledge of hidden things" and "growing larger or smaller at will."

The Grant Park study is the third and longest of a series—following two others of one week's duration. And in 1978 Maharishi reportedly sent goodwill meditators to transmit pure vibes in five world trouble spots, including Iran and Nicaragua.

*"Literary intellectuals at one pole—
at the other, scientists. . . .
Between the two a gulf of mutual
incomprehension."
—Charles Percy Snow*

PREMATURE THAW When the Cryonics Society, of Los Angeles, California, was founded in 1965, it offered to freeze the newly dead in liquid nitrogen for the sum of $20,000. Sustained at $-320\,°F$, the corpses would be too cold to decay, the society said, and, in the future, if a cure were to be found for the disease that the deceased died of, the body could be reanimated and healed.

Recently, however, the booming cryonics enterprise was dealt a death blow. A California Superior Court has awarded three families nearly $1 million damages against the Cryonics Society for gross negligence and fraud. The families involved sued after finding that the bodies of relatives in cryonic suspension had thawed and decomposed. The California company is no longer extant, but other cryonics organizations continue doing business, at least for a while.

Public opinion varies regarding the future of cryonics in this country. Robert Ettinger, known as the father of cryonics, considers the incident a typical accident that could have happened in any industry. Others disagree and consider the cryonics business a ripoff. John Gill, secretary of the California Cemetery Board, says, "Cryonics is obviously consumer fraud. They can't do what they promise—that is, bring you back from the dead." In summing up, he says, "There is no known technology available today to prove that cryonics works and is a viable method of preservation."

HYPNOREADING High-school students in a darkened Los Angeles classroom shut their eyes and listen to the lulling voice of a hypnotist. He firmly reassures them that they will learn to read faster with each passing day, and as he speaks, they visualize the words of an imaginary book gliding past their field of vision.

When the session is finished, they are given homework: exercises to quicken their eye movement and daily self-hypnosis sessions in which they repeat the instructor's suggestions again and again.

241

"Among theories finally rejected by scientists in the early Nineteenth Centry [was the belief] that mice could be produced by enclosing a hunk of cheese and some old rags in a hat box."
            —*E. C. Large*
            in The Advance of the Fungi

The students are learning hypnoreading, an offbeat speed-reading method developed by Live and Learn, a nonprofit educational foundation in Sherman Oaks, California. According to program directors Michael Lilienfeld and Steven Snyder, successful students must practice under hypnotic trance daily for several weeks. While the program requires a strict regimen, it does not take the pleasure out of reading—a complaint often leveled at more conventional speed-reading systems. "This is because improved skills acquired during practice sessions are automatically used during the normal waking state," Lilienfeld says, "without the reader's having to make any conscious effort."

Hypnoreading is an extension of Live and Learn's "alpha learning" program, in which students are trained to change their brain rhythms from the beta waves of normal waking hours to the slower alpha waves produced during such activities as daydreaming, watching television, and listening to soothing music. Studies have shown that gifted children have slower brain rhythms and tend to be in the alpha state more frequently, Lilienfeld explains. "We believe that the slower the brain's rhythm, the more receptive the neural passages are to receiving information."

For the past three years Live and Learn has been retained by the Los Angeles School District to teach alpha learning and hypnoreading to students and school counselors. Participants say that they have improved information recall and that they read three to five times faster than before.

ALIEN ABDUCTIONS Aphrodite Clamar has hypnotized 19 people who claim they were kidnapped by extraterrestrial beings, but she still can't determine whether these experiences were genuine.

"I'm perplexed," the forty-two-year-old New York City psychotherapist admits. "Most fantasies are pleasurable, yet the tales these people told were terrifying. I know it was a real experience, at least in the minds of the people who had it. I don't believe they were feigning."

A man who babbled
bad Russian in his sleep
was forced to
get an unlisted phone number.

Clamar says that 15 of the 19 subjects, all rounded up for her by UFO buff Budd Hopkins, told strikingly similar stories. Most claimed they were abducted, sometimes twice, for periods of up to two hours, during which time they were operated on by "smallish humanoids," with "puttylike" skin and "round, metallic" eyes. During the surgery, usually performed under eye-straining white light, tissue was extracted from their bodies, presumably for scientific evaluation. The subjects also claimed that the aliens wore "thick-textured, gray one-piece jumpsuits."

The 19 "abductees" included a lawyer, a stockbroker, a teacher, a television newsman, and an actor, Clamar says. "None," she adds, "were marginal souls."

One of the psychotherapist's latest findings: "Males, especially teen-aged males, did not fare as well as females in integrating and accepting the experience of extraterrestrial abduction in their daily lives." Typically, male subjects viewed their abduction with a "sense of inadequacy, a poor self-image, and discomfort," but females "appeared better able to discuss their experiences and find a niche for them."

RUSSIAN SLEEPTALK Wilma Sutherland, of Mesa, Arizona, was roused from her dreams whenever her husband, Gene, talked in his sleep. Groggily perking her ears, she would make out words and phrases and, satisfied that nothing was amiss, would immediately fall asleep again.

One night, though, the babbling that woke her was different—more excited and agitated, and rife with sounds like "ski" and "vich" repeated in a thick, unfamiliar accent. None of it was intelligible.

Somewhat unnerved, Wilma got a tape recorder to capture the gibberish. When she played the tape back to Gene, he had no idea what the sounds meant or what had triggered them. Nonetheless, because they reminded her of Russian, she called the foreign language department at Arizona State University.

*"There's a hell of a good universe
next door, let's go."*
                    —e.e. cummings

Hearing the 40-minute tape, Professor Lee Croft recognized eight or nine Russian terms, including "Pjanj," which means "a drunk," "prostividno," or "excuse me, it's evident," and "on byl i," translated as "He was, and."

Croft learned that Gene Sutherland's one previous experience with the Russian language had occurred when he was a serviceman in Germany at the end of World War II, at the time the U.S. Army united with the Red Army at the Elbe River. As far as Sutherland remembered, none of the Russians he met taught him any words in their language. Croft surmised, however, that the experience had somehow left a deep impression on Sutherland, etching the Russian phrases into his subconscious.

The Sutherlands, meanwhile, were fascinated by the incident, telling the story to all their friends and neighbors. When the local newspaper got wind of the Russian dreams, it ran an article together with a photo showing Gene under the covers. Suddenly droves of reporters, religious zealots, and even "parapsychological authorities" were beating a path to the Sutherlands' door. They urged the couple to try hypnosis and other psychological ploys to find out Gene's secret and told them that the ability to talk Russian was caused by everything from reincarnation to demonic possession. Gene was even accused of being an unwitting Soviet tool, contacted—and manipulated—by the Russians through telepathy.

Whatever the explanation, the Sutherlands don't care much anymore. Hounded and exhausted, they have gotten an unlisted telephone number.

PLACEBO HOROSCOPES Are horoscopes that have been carefully prepared for a specific person more accurate than "placebo" horoscopes intended for someone else? Absolutely not, says Douglas P. Lackey, a philosophy professor at Baruch College, in New York City.

Lackey used a standard astrology manual to prepare personal horoscopes for 38 Baruch students. Then he gave each student a copy of his own horoscope, as well as another one selected at random. Each horoscope

Lancelot: the world's
first unicorn? Or merely the
product of some
careful horn grafting?

contained approximately 20 paragraphs. Students were requested to rate the accuracy of each.

According to Lackey, students felt that their own horoscope and the placebo were more or less equal in accuracy: Nineteen said that their own was more accurate, eighteen said that the placebo was more accurate, and one rated the two charts dead even.

UNICORN Legend has it that the last living unicorns were reported in the Near East in 1503, by one Luigi Vartoman, of Bologna, Italy, who claims he saw two unicorns while at the palace of the sultan of Mecca.

Now legend has allegedly become life. Two naturalists residing in Mendocino County, California, claim they have bred a unique animal whose fiery head and flowing mane are capped by a single horn growing from the middle of the brow. "Lancelot," as he is called affectionately by his breeders, is possibly the first living unicorn in modern times.

Born to an Angora goat and an as-yet-unidentified animal (the owners won't discuss this part), the unicorn stands 2.5 feet tall, is 3.5 feet long, weighs 75 pounds, has cloven hooves, and is pearly white. "Creating Lancelot was the result of duplicating past research through interbreeding," say owners Morning Glory (a woman) and Otter G'Zell (a man).

The breeders claim that the secret of their accomplishment came from many hours of painstaking examination of medieval tapestries and transcripts. They now plan to patent the unicorn process.

Lancelot's detractors, however, claim he is an ordinary goat whose horn buds were removed, then cut in two. Half of each horn, say critics, was discarded, and the remaining halves placed together in a small wound in Lancelot's forehead.

SUBJECT INDEX

CHAPTER 1: HEALTH & MEDICINE   3
  Brown-Fat Diet   4
  Barley Juice, Cancer, and Aging   6
  Interferon Lotions   7
  Killer Amoebas   7
  Crib-Death Cure   8
  Emotion and Sudden Death   9
  Fetal Kicks   9
  Computerized M.D.   10
  Hair Diagnosis   11
  Bowel Bullets   12
  Lightning Recovery   12
  Jogging Pigs   13
  Secret Milk Ingredients   14
  Caffeine Confusion   14
  Microwave Death   15
  Better Pregnancy Test   16
  Video Therapy   16
  Stone Age Scalpels   17
  Listening to Bones   18
  Nausea Straps   19
  Dusty Nuns   20
  Prescription Errors   22
  Rectal Laser   22
  Sharks' Cancer Secret   23
  Spine Medicine   24
  Bloodsuckers   25
  Secret of the Pygmies   25
  Binge Disorder   26
  The Ouch Room   27
  Super Scan   29
  Icy Operations   30
  New Runner's Problem   31
  VDT Hazards   32
  Randy Yeasts   33
  Carbon Tendons   34
  Blood Pills   35
  Allergy Treatment   35
  Personal Heart Computer   36
  Teen Malnutrition   37
  Shocking Poison   37
  Jello Lives   38
  Smoking Babies   39
  Bloodletting for Health   40
  Cancer Computer   40
  Leech Nerves   41
  Eye Calisthenics   42

  Protein Slices   42
  Computer Surgery   43
  New Skull   44
CHAPTER 2: ANIMALS & WILDLIFE   45
  Animal Love   46
  Reformed Killer Bees   47
  Nuclear Gophers   48
  Monkey Orgasm   48
  Ascent of Man   50
  Mice With Antlers   50
  Toasted Termites   51
  Mystery Bird   52
  Frog Crisis   53
  Monkeys for the Handicapped   53
  Transvestite Flies   54
  Pets and Blood Pressure   55
  Animal Incest   56
  Bird Brains   57
  Whither the Flamingo?   58
  Whale Love Songs   60
  Suspended Animation   60
  Insect Brain Transplants   61
  Social Lubricants   62
  Twisting Birds   63
  Polar Bear Energy   64
  Friendly Cats   65
  Wheat Aphrodisiac   66
CHAPTER 3: STARS & SPACE   67
  Viruses From Outer Space   68
  Space Goods   69
  A Star Named Sue   69
  Death on Mars   71
  American Stonehenge   72
  Shrinking Sun   72
  Goats in Space   73
  Planet Trails   74
  Space-Surgery Sack   75
  Hot Real Estate   75
  Comet Shock   76
  Space Van   77
  $3 Million Toilet   78
  Space Age Bra   79
  Black Holes Rescued   79
  Moon Billboards   81
  Poor Man's Communications
    Satellite   82
  Asteroid Neighbors   83
  Asteroid Bombs   84

Gravitational Lens	85	Ocean Energy	127
Space Robots	86	Prospecting Without Drills	128
Drugs in Space	87	Solar Mobile Home	129
Space Age Lube	87	Energy Mountain	129
Diamonds in Outer Space	88	Solar Airplanes	131
Gravity Waves Found at Last	89	Hot Rocks	132
Martian Meteors	90	Leaves of Gas	133
		Energy Tax Shelters	133
CHAPTER 4:		Hot Jello	134
BEHAVIOR AND THE MIND	91	Energy From Dumps	135
Hotdogger or Hamburger?	92	Garbage-Powered Cars	135
Women in Love	92		
War-Neurosis Drug	93	CHAPTER 6: TECHNOLOGY	137
Einstein's Brain	94	Robot Evolution	138
Lucid Dreaming	96	The Airmobile	139
Separate Checks	97	Synmetals	139
Sexual Fantasies	98	Electric Plastic	140
Mammary Madness	99	Blindsight	141
Mating Game	100	Super Sperm	142
Born Fighters	101	Red-Tape Measures	143
Violent Pornography	101	EyeDentity	144
Grass Intelligence	102	Ultrasonic Steak	146
Easy Believers	103	Chemometrics	146
Opening Lines	104	Electric Roads	147
Snoring Cure	105	Lasers vs. Baldness?	148
Sporting Violence	106	Recombinant Diarrhea	148
Odd Couples	106	Long-Life Banana	149
Plastic Brains	107	Vegetable Evolution	151
Menarche Myth	108	Silicon Cyrus	152
Sleep/Wake Biofeedback	109	Super-Pressure	
Emotional Alarm Clock	109	Superconductors	153
Shock Therapy	111	Air Force Robots	154
Half a Brain	112	Stealth Plane	154
Hostile Shortness	113	Chunnel Update	155
Damp Genesis	114	Amazing Laser Facts	156
Forgettable Faces	115	Miracle Materials	157
Marijuana-Heroin Link	117	Japan's Supercomputer	157
The Myth of Mother Love	117	Moving Building	158
Our Third Eye	118	Chrome Independence	159
Life in the Womb	119	Murdered Computers	160
Dominican Switch	120	Hydrogen Factories	162
Potential Rapists	121	Laser Photochemistry	163
Radical Autism Therapy	121	Metallic Glass	164
Athletic Brain	122	Corny Plastic	164
CHAPTER 5: ENERGY	123	CHAPTER 7: PHENOMENA	165
Energy Train	124	Mouth Birth	166
Fuel-Saving Wings	124	Book Death	166
Italian Tower of Power	125	Green Hair	167
Plant Petrol	126	Tongue Twister	168

Human Carrot 169
Urban Legends 169
Primal Growth 170
Whistling Air Crashes 171
Mermen and Mermaids 172
Ancient Recordings 173
Future Faces 175
TV Census 175
Noisy Vegetables 177
Magnetic People 177
The Bedford Pyramid 178
Whistling Ears 179
Fossil Music 180
Future Millionaires 180
Coincidence 180
Electric Flying Carpets 181
Surprise Sight 182
Plastic Bees 183
Tasty Genes 184
Plant Stress 185
Volcano Deaths 186
Sniffing Out Limburger's Scent 187
Changing Accents 187

CHAPTER 8:
INVENTIONS & INNOVATIONS 189
Foul Mouthwash 190
Rock and Roll Hot Pants 190
Conspicuous Drunks 191
Speech for the Deaf 192
Cat-Flap Bomb 193
Rainy-Day Genes 194
Talking Checkout 194
Car Finder 195
New Suits for G-Men 196
Movable Mailbox 197
Talking Thermometer 197
Buy or Rent? 199
Spying on the Wright Brothers 199
Double Vision 200
Future Radio 201
Holocaust Hive 202
Wine for Teetotalers 203
Superskates 204
Custom TV Advertising 204
Reinventing the Oxcart 205
Aqua Pura Mobile 207
Cell Cinema 207
Do-It-Yourself Weather
  Forecasting 208

Maps in Flight 209
Simulated Baseball 209
Diet Fork 210
Talking Briefcases 211
Rippling Seat 212
Fast Talk 213
Seeing Through Smoke 213
Artificial Ear 214
Nonreversing Mirror 215
Laser Spots Stolen
  Gems 215

CHAPTER 9: ENVIRONMENT 217
Soviet Chill 218
Survival Condo 218
Mississippi Sea 219
Burmese Eco-Engineering 220
Spoiled Parks 220
Poison Parsnips 221
Snakescam 222
Honey Crisis 223
Dangerous Fluorescents 224
Tree Filters 225
Radiation Shield 226
Russian Invaders 227
Tropical Rip-Off 228
Monsoon Madness 228
Pollution Canary 229
It's Still Hard to Change
  the Weather 230

CHAPTER 10: THE PARANORMAL
(AND OTHER WEIRD THINGS) 231
Human Combustion 232
Dogu Space Suits 232
Fang Count 233
Loch Ness Worms 234
Nostradamus Interpreted 234
Haunted Personalities 236
China's Armpit Savants 236
Remembering Birth 237
Sole on Fire 238
UFO Detectors 239
OM Swat Team 240
Premature Thaw 241
Hypnoreading 241
Alien Abductions 242
Russian Sleeptalk 243
Placebo Horoscopes 244
Unicorn 245